U0163704

# A Guide to
# Time Travel
# Through Its Past of
# Xi'an Jiaotong University

西安交通大学
穿越指南

李　重　程洪莉　　主编

西安交通大学出版社
XI'AN JIAOTONG UNIVERSITY PRESS

国 家 一 级 出 版 社
全国百佳图书出版单位

## 图书在版编目（CIP）数据

西安交通大学穿越指南：汉、英／李重，程洪莉 主编 .—
西安：西安交通大学出版社，2021.6

ISBN 978-7-5605-9107-0

Ⅰ．①西… Ⅱ．①李… ②程… Ⅲ．①西安交通大
学 - 教育建筑 - 指南 - 汉、英②西安交通大学 - 校园 - 景观 -
指南 - 汉、英 Ⅳ．① TU244.3-62

中国版本图书馆 CIP 数据核字（2020）第 143304 号

西安交通大学穿越指南

XI'AN JIAOTONG DAXUE CHUANYUE ZHINAN

| | | |
|---|---|---|
| **责任编辑** | 王斌会 蔡乐芊 | |
| **责任校对** | 邓　瑞 | |
| **出版发行** | 西安交通大学出版社 | |
| | （西安市兴庆南路 1 号 邮政编码 710048） | |
| **网　　址** | http://www.xjtupress.com | |
| **电　　话** | （029）82668357 82667874（市场营销中心） | |
| | （029）82668315（总编办） | |
| **传　　真** | （029）82668280 | |
| **印　　刷** | 西安五星印刷有限公司 | |

**开　　本** 710mm×1000mm　1/16　　**印张** 11.25　**字数** 218 千字
**版次印次** 2021 年 6 月第 1 版　　2021 年 6 月第 1 次印刷
**书　　号** ISBN 978-7-5605-9107-0
**定　　价** 78.00 元

如发现印装质量问题，请与本社市场营销中心联系、调换。
订购热线：（029）82665248　　（029）82665249
投稿热线：（029）82668525

**《西安交通大学穿越指南》编委会**

总策划： 成 进
主 编： 李 重 程洪莉
副主编： 杨澜涛 李 明 董 琪
编 委： 吕 青 许佳辉 宁春明 李一鸣 王 劲
　　　　李 慧 朱建欣 吴 丹 谢霞宇 刘鸿翔
　　　　戴毓珩 曲建利
翻 译： 楚建伟 贺子煜 史家琪 李 玮 曹 琰
设 计： 董 琪 刘憬晗
出 品： 西安交通大学党委宣传部
支持单位：西安交通大学档案馆、博物馆
　　　　　西安交通大学人文社会科学学院
　　　　　西安交通大学翻译中心
　　　　　西安交通大学校园规划与基本建设管理中心

# 交通大学总体平面佈置图

| 编號 | 名 称 | 編號 | 名 称 |
|---|---|---|---|
| 1 | 中心大楼 | 21 | 運動場地 |
| 2 | 電制大楼 | 22 | 丁种宿舍 |
| 3 | 動力大楼 | 23 | 混合住宅 |
| 4 | 探礦大楼 | 24 | 車 庫 |
| 5 | 机制大楼 | 25 | 風雨操场 |
| 6 | 圖書館 | 26 | 鍋炉房 |
| 7 | 實習工廠 | 27 | 庫 房 |
| 8 | 診疗室 | 28 | 高壓試驗室 |
| 9 | 工會俱樂部 | 29 | 贱品倉庫 |
| 10 | 雜 屋 | 30 | 扇工戎驗坊 |
| 11 | 大禮堂 | 31 | 第二食堂 |
| 12 | 學生食堂 | 32 | 單身宿舍 |
| 13 | 浴 室 | 33 | 高級住宅 |
| 14 | 學生宿舍 | 34 | 拆成鍋炉房 |
| 15 | 福 利 | 35 | 臨時房屋 |
| 16 | 托兒所 | 36 | 中央部電所 |
| 17 | 小 學 | | |
| 18 | 幼兒園 | | |
| 19 | 甲种宿舍 | | |
| 20 | 乙种宿舍 (回地201号) | | |

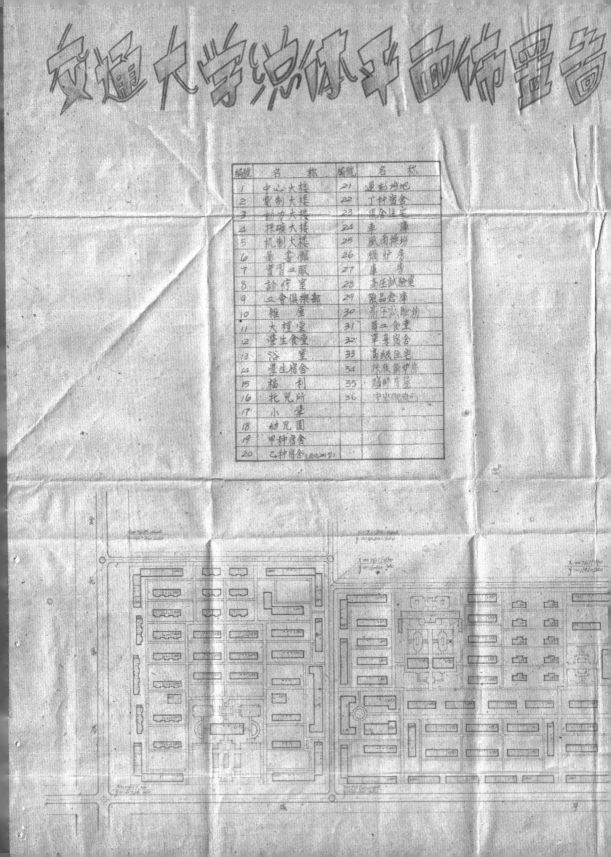

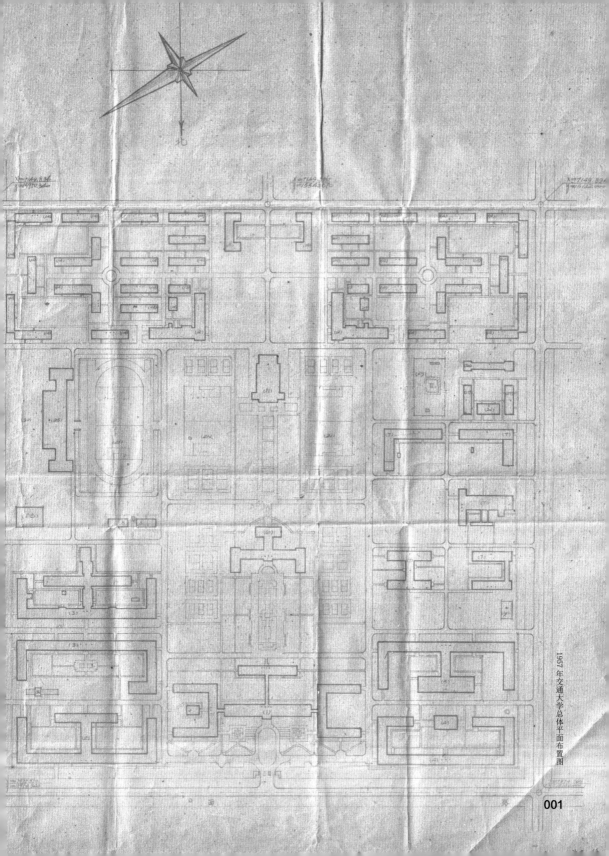

1957年交通大学总体平面布置图

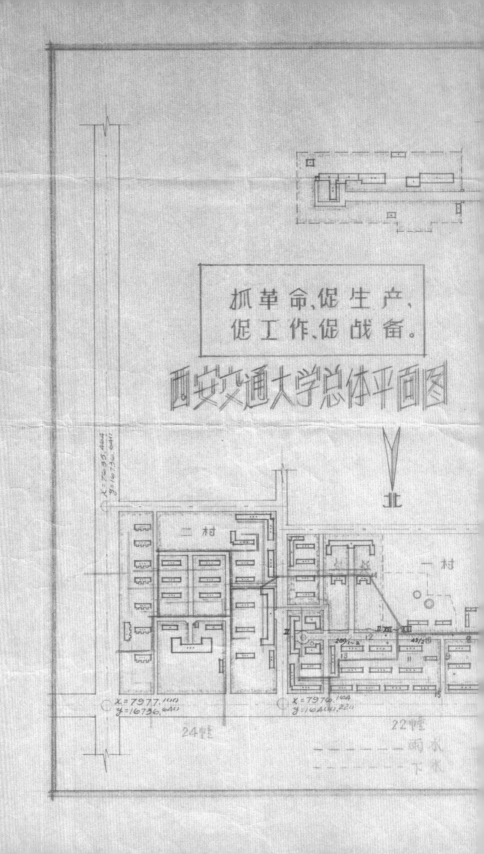

抓革命、促生产、
促工作、促战备。

西安交通大学总体平面图

北

二村

一村

24甲?

22甲?
——————雨水
————下水

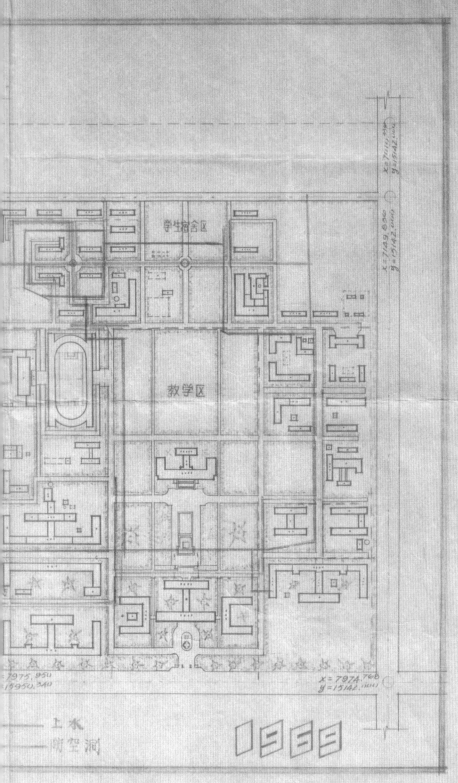

学生宿舍区

教学区

x=7149.650
y=15142.000

x=7011.550
y=15142.000

x=7974.768
y=15142.1004

7975.950
15950.340

上水
明堑洞

1969

西安交通大学总体平面图  1/5000

# 西安交通大

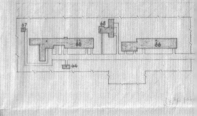

| 编号 | 名称 | 编号 | 名称 | 编号 | 名称 | 编号 | 名称 | 编号 | 名称 |
|---|---|---|---|---|---|---|---|---|---|
| 1 | 中心大楼 | 19 | 爆炸试验室 | 37 | 游泳池 | 55 | 470试验室 | | |
| 2 | 电制大楼 | 20 | 多冲试验室 | 38 | 凤丽操场 | 56 | 贮氢贮室 | | |
| 3 | 动力大楼(乙) | 21 | 实习工厂(丁) | 39 | 更衣室 | 57 | 小汽车房 | | |
| 4 | 动力大楼(甲) | 22 | 实习工厂(戍) | 40 | 动态模拟试验室 | 58 | 托兒所 | | |
| 5 | 机制大楼 | 23 | 设备仓库 | 41 | 水电组 | 59 | 幼兒园 | | |
| 6 | 计算机站 | 24 | 木工间 | 42 | 学生宿舍 | 60 | 敎工食堂 | | |
| 7 | 无线电工厂 | 25 | 洗车库 | 43 | 学生浴室 | 61 | 浴室锅炉房 | | |
| 8 | 电镀车间 | 26 | 水池 | 44 | 锅炉房 | 62 | 單身宿舍 | | |
| 9 | 学工车间 | 27 | 圖书馆 | 45 | 临时食堂 | 63 | 粮站 | | |
| 10 | 工务楼 | 28 | 印刷厂 | 46 | 学生第一食堂 | 64 | 邮局 | | |
| 11 | 北区锅炉房 | 29 | 制冷试验室 | 47 | 蓄水池 | 65 | 服装社 | | |
| 12 | 实习工厂(甲) | 30 | 动力机械厂 | 48 | 水泵房 | 66 | 工程物理试验室 | | |
| 13 | 实习工厂(乙) | 31 | 锅炉车间 | 49 | 灯光球场 | 67 | 庆库 | | |
| 14 | 大锻车间 | 32 | 发电机房 | 50 | 治学生宿舍 | 68 | 液态金属鈉 | | |
| 15 | 锻机厂 | 33 | 东区锅炉房 | 51 | 学生第二食堂 | 69 | 盐景坊 | | |
| 16 | 实习工厂(两) | 34 | 高压试验室 | 52 | 危险品仓库 | 70 | 校门 | | |
| 17 | 手锻工厂 | 35 | 起动试验室 | 53 | 水塔 | | | | |
| 18 | 金属结构厂 | 36 | 碟修泵 | 54 | 敎工住宅 | | | | |

永久性建筑物

临时性建筑物

拟建建筑物

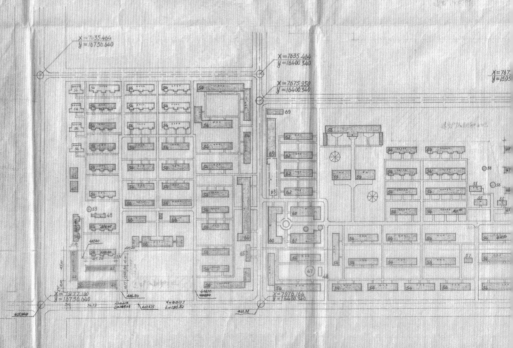

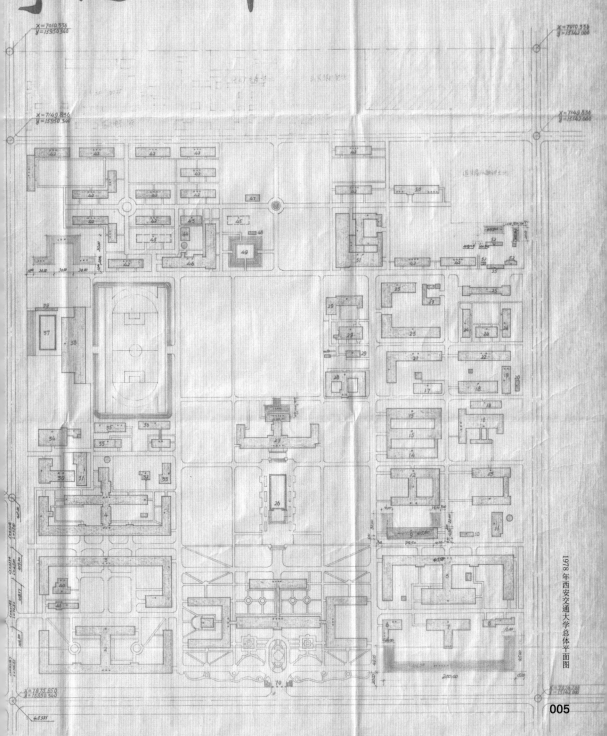

## 学总体平面图 1:2000

1978 年西安交通大学总体平面图

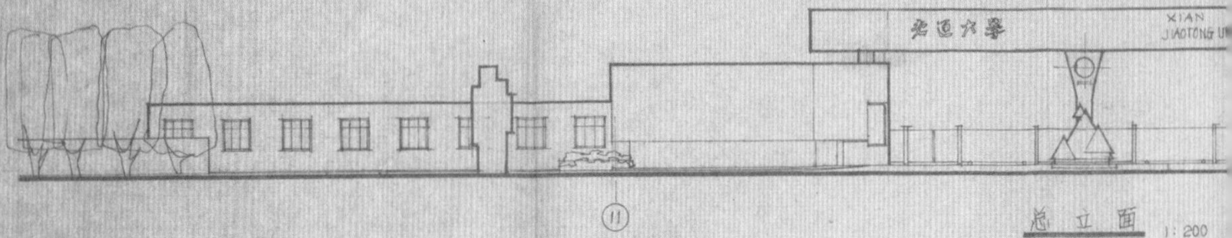

交通大学

⑪

总立面　1：200

# 说　明

1. 本设计是对原校门设计的修改，力求便于施工，轴线沿用原设计轴线，施工图参看原设计施工图。

2. ±0.00标高和原设计相同。

3. 浮雕和雕塑造型应由创作者完成。（浮雕坪类面喷铜）。

4. 水暖电消的图面和原设计相同。

5. 本设计提供的绿化方案为椭圆绿地及碎石路面设计。除选用行道树2列，其余植物可随种植高大细叶，以开阔场地增大观瞻。椭圆中设置大型彩色金属抽象雕塑，本构图为日本主题的草坪草坪，回字绿篱，西面乔草，观瞻地面，绿化经费另列考虑。

坪面做法：

坪1. 浮雕坪面：300厚 300#白水泥石英砂白石子钢筋混凝土坪
部筋：竖向 Φ10@200.
水平向 Φ10@200.

坪2. 玻璃锦砖坪面：15厚 1：3水泥砂浆找平
1：1水泥砂浆粘贴
白色陶瓷马赛克（防水泥粘贴）

坪3. 斩假石坪面：15厚 1：3水泥砂浆找平
10厚 1：1.25水泥石屑浆面层，斩假

坪4. 圆坪坪面：15厚 1：0.5：5白混合砂浆找平
红褐放原色
10.6白色涂料防白两道

地面做法：

地1. 室内地面：素土夯实
100厚 3：7灰土夯实
80厚 100#混凝土垫层
20厚 1：3水泥砂浆面层（加氯化钙5%）

地2. 室内地面（预制磨原砖）停车围圈际线 部分：素土夯实
80厚 100#混凝土垫层
20厚 1：3水泥砂浆粘贴
20彩瓷釉面（水拼防浆划缝）花纹红色.

地3. 路面：素土夯实
300厚碎石路层夯实
100厚新青石热铺,压实机压平

8. 屋面防水：（挡板门雨篷，其余同设计）
素注防水屋面：1：8水泥炉渣找坡，最薄处15厚
1：3水泥砂浆找平
冷底子油一道
二毡三油防水层
绿豆砂保护层

9. 油漆：
所有钢件的油漆：防锈漆三道
（所用混合漆三道，用于DM-1,DM-2）
调合漆三道
门窗塑围白板：木色清漆
反 XM-2
XM-3、XC-2: 同原设计.

10. 门窗明细表

| 门窗编号 | 洞口尺寸 | | 选用图集 |
|---|---|---|---|
| DM-1.2 | | | 详见建施－8 |
| XM-1.1 | 2700×1500 | 4.5 | 详见建施－7 |
| XM-2.9 | 2700×2060 | 2.12 | 同上 原设计选用M＋玻璃可改 |
| XM-3.9 | 1800×800 | 1.64 | |
| XC-1.1 | 2100×4200 | 8.32 | 详见建施－7 |
| XC-2.1 | 600×600 | 0.36#? | |

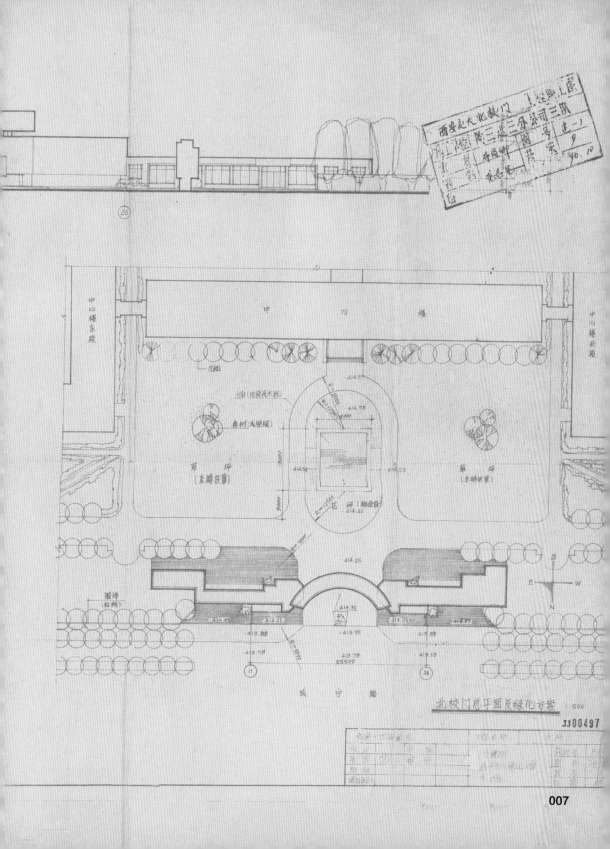

北校门总平面及绿化方案  1:500

JJ00497

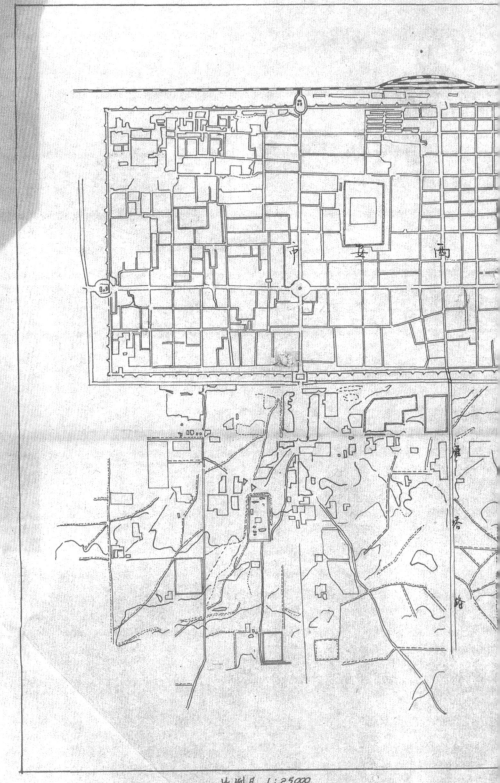

比例尺 1:25000

交大校址地位圖

西迁初期交大校址地位图

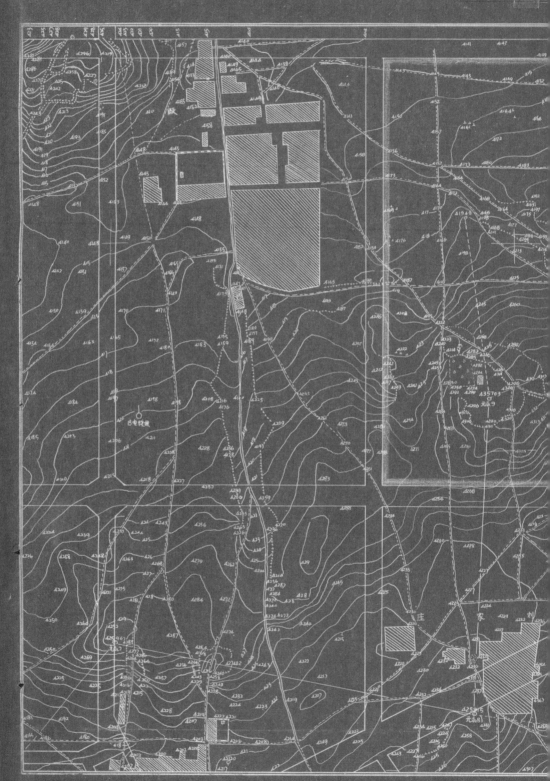

比例 1:5000

1956年交大校址地形图

上：1957年中心运动场 下：20世纪70年代中心运动场

2000 年中区食堂

改革开放初期校园全景鸟瞰图（模型）

左上：1957-1958 学年交通大学开学典礼在草棚大礼堂举行　左下：1955 年 5 月彭康校长（左四）一行在西安选址　左中：西迁初期西安新校的员工宿舍外景

本报讯　陕西省高校优秀教学成果表彰奖励大会于4月11日在人民大厦礼堂举行。省政协主席周雅光、副省长孙达人出席大会颁奖。孙达人到省长在评话中希望大师广大师生参加到高校之不易的安定团结的局面，树立稳定压倒一切的思想，积极维护国家和社会稳定，希望广大教师教育学生全面发展，改进和加强思想政治工作，把育人的基础落在实际、改进教学方法、提高教学质量，取得更大的成果。

会上四月份连续被误工参观厂矿企业，以接触社会、接触实际，四月十四日部分青年教职工在西北国棉五厂参观。　　王敏实摄

根据学校全面发展，德智体全面发展，还有一部分同学是出于不满足政治理论课上对马列主义毛泽东思想的初步介绍，希望更了他所在教研室坚持培养青人的经验。特别是毛泽东思想同中国革命具有关著作，针对如何树立正确的世界观开展讨论，倡控系的学习小组则结合学习《青年运动的方向》等有关著作，学习小组则侧重面有所不同。网机、物理系机械学系的学习小组主要全心全意为人民服务的

《共产党宣言》、《费尔巴哈和德国古典哲学的终结》等马恩著作。在学习毛泽东著作时，各学习小组侧重面有所不同。也有部分学习小组是同学们自发组织的，有的是在教师的指导下学习

据统计，现全校学生中已成立马列和毛泽东著作学习小组27个，成员约380人，遍布全校绝大部分系。这些学习毛泽东著作为主要内容，也有的是全校立的，也有的是组合起来组织学习小组的，参加学习的学生包括各个年级的同学。其中大部分学习小组以学

## 大学生读马列渐成风气

从上学期来到现在，我校学习马列和毛泽东著作，阅读介绍马东著作，领袖生平事迹的书籍、文章，给充满劲勃机的校园注入了一股新的热流。

"知无不言，言无不尽"，气氛限受欢迎。限受益。"赴京出席第七届全国政协二次会议归来的陈学俊，4月11日在402会议室里兴致地起自己参观来的感受。

陈学俊教授说，"两会"在民主、求实、团结、奋进的基础上充分发扬了社会主义民主；二是与会同志坚持四项基本原则的立场不动摇。具有两个明显的特点，一是在民非常关心国家大局的稳定，体现出了大家不心协力团结奋斗的精神，三是比

西安交通大学主办　　第229期　　1990年5月5日
陕西省内部报刊统一刊号　SXⅡ——0134　共4版

## 国家教委召开直属高校后勤工作研讨会

（李新春）

## 第三届科技文化艺术节闭幕

### 史校长题词：生机勃勃 奋发向上

本报讯 4月18日在学生活动中心开幕的《辉煌的1989年》大科技成果，拉开了我校第三届"科技、文化、艺术节"的序幕。

这届"科技、文化、艺术、学生会、研究生会联合举办的、史校长为本次活动的主题"生机勃勃、奋发向上"揭示了这次科技文化艺术节主题。

## 关于通报表彰魏文元、胡卫、田毅平、杨牧的决定

附註：

一．每人除帶被褥、衣物、日用品等外，還應攜帶下列物件：

（一）准考證、本通知書；

（二）學歷證件；

（三）戶口遷移證、油糧供應轉移證；

（四）在職青年須攜帶原機關離職證明文件；

（五）凡家庭經濟困難欲申請人民助學金者，原在中學有助學金者，須攜帶原校助學金證明，並須一律按我校所寄「申請人民助學金證明書」之要求填寫，交當地地區人民委員會，或家長所在國家機關，企業予以證明，由該證明機關郵寄我校。

（六）一寸正面半身脫帽照片八張；

（七）繳交書籍費約十二元；

二．報到地點：西安市緯十街經七路。

三．錄取新生如不及如期來校報到應在接到通知書後即來信向本校教務處申明理由辦理請假手續。

四．各考區錄取新生如何去西安另行附告

# 通知書

沈生華同學：

根據國家建設的需要，參照你的志願和條件，經過我們的審查，你在這次高等學校招生考試中，已正式被錄取分配入交通大學□□類學習。

為了爭取早日入學學習，請你在九月一二兩日內，憑本通知書並攜帶後列應帶物件赴校報到（逾期不報到，不請假者，即取消入學資格）。

赴校路費自行籌措；如你家境確實窮苦無力籌足路費，可按照原考區招生機構規定之新生赴校報到路費補助辦法，請求補助。

親愛的同學！我們偉大祖國已經開始了社會主義經濟建設，國家迫切地需要各類建設幹部。希望你早日作好準備，愉快地走上學習崗位，接受祖國交給你的學習任務，爭取成為祖國建設的合格人才，建成社會主義社會而努力。

交通大學

鄂州市第一中学：

你校应届毕业生

今年高考成绩优异，已

老师们辛勤教导的结晶

今后加强联系，共同为

西安

一九

报

奏国兵同学

取在我校学习，这是贵校

将向你们报喜。让我们两校

养四化建设人才而努力。

通大学

五年九月

1985年西安交通大学发鄂州市第一中学喜报

# 証 明 書

(61)交总字第0009720号

　　兹有我校 学生 胡岳　同志在
教工

暑假期間同意其离校返乡50天（由

7月3日起至8月29日止）

发给粮票伍拾斤，停止其他供应，

請你处按当地的供应标准供应蔬菜

和其他付食品为荷

　　此致

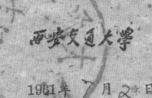

西安交通大学

1961年7月2日

1961 年粮票证明

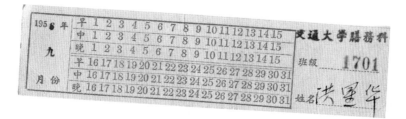

20 世纪 50 年代 -90 年代的粮票、菜票

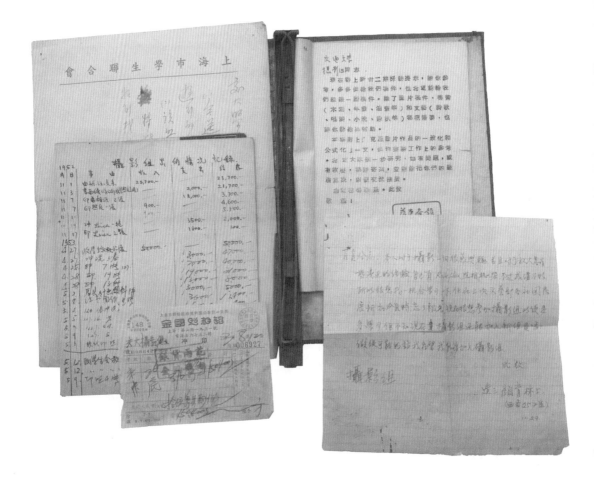

20 世纪 50 年代学生宣传摄影活动记录

# 引言

这里是咸宁西路 28 号，西安交通大学兴庆校区。

"兴庆"之名，源于学校北门对面即为兴庆宫——与太极宫、大明宫并称为唐长安城"三大内"。而此处校园便位于汉代上林苑范围内，唐朝时道政、常乐二坊所在。

俯览校园，瓜剖棋布，尽显对称之美；碧树丹甍，南北错落有致。尤其那两条郁郁葱葱的梧桐大道，夏日青翠欲滴，秋日金黄璀璨，四时风景不同，但走在树荫下却同样静心。

这里的学子赠母校一个飘逸昵称——"仙交大"。

"仙"不止因为校园优美，还因其蕴藏的气象万千。穿越历史的星空，你是否依稀瞥见，汉武帝在上林苑策马扬鞭的矫健身影，唐玄宗和杨贵妃共赏《霓裳羽衣》的缠绵悱恻，李白挥就"云想衣裳花想容"的热情豪放，白居易写下《养竹记》的耿介直爽……

"仙"也不止因为文化风流，更因一群超凡脱俗的大师名家。走过时光的长河，你似乎可以听见，"中国电机之父"钟兆琳讲授"感应电机"的行云流水，热力工程先驱陈大燮剖析"熵"的透彻清晰，能源动力工程科学家陈学俊歌唱《革命人永远是年轻》的豪迈慷慨，机械故障诊断领域奠基人屈梁生创建全息谱所经历的发动机轰鸣……

这片土地，辉煌过也黯淡过，伟大过也沉寂过，有过马蹄声如奔雷，也曾麦浪滚滚闪金光。它见证了汉武帝的雄才大略，盛唐的繁华似锦，太学的博士从这里走过，狂傲的诗人从这里走过，无数古来的贤者从这里走过……它见证了三秦大地的沧桑巨变，一座高等学府的拔地而起，西迁的先行者来到这里，五湖四海的青年才俊来到这里，西部开发的拓梦者来到这里……

来吧，一起开始穿越之旅吧！一起寻觅千年前的云兴霞蔚，一起回味交通大学西迁时的波澜壮阔；一起与吴道子、韩干共绘丹青，一起与西迁老教授畅谈科学；一起品尝这沉淀愈久愈香醇的琼浆玉液，一起流连这动人心扉的交大校园。

仙交大，愿你不虚此行。

# Introduction

This is Xingqing Campus of Xi'an Jiaotong University (XJTU), No.28, Xianning West Road, Xi'an, Shaanxi Province.

The name of "Xingqing" derives from Xingqing Palace, one of the three major palaces in Chang'an City of the Tang dynasty just located across the street to the north of XJTU. The other two major palaces then were Daming Palace and Taiji Palace. The present campus used to be Shanglin Imperial Park in the Han dynasty, and Daozheng Fang and Changle Fang in the Tang dynasty.

Looking from above, all buildings on campus are rowed symmetrically with trees and red roofs located in a row running north to south. The lush East and West Plane Tree Roads, in particular, are charming with a wild profusion of trees, green in summer and golden in autumn. Although the scenery is various in different seasons, one would always feel calm and quiet walking in the shade of the trees.

The students nicknamed their *alma mater* —— *Xian Jiaoda*, the Heavenly Palace.

To use "Heavenly Palace" in Chinese to describe the university is not only because of its beautiful scenery on campus but also the historical events on the site. Travelling back, you could see Emperor Wu of the Western Han dynasty riding in the Shanglin Imperial Park, Emperor Xuanzong of the Tang dynasty and his imperial consort, Yang Yuhuan exceedingly sentimental about the "Rainbow Skirt and Feather Garment Song", Li Bai, a famous poet in the Tang dynasty passionately composing his poem "A Song Of Pure Happiness I", and Bai Juyi's, a great poet of the Tang Dynasty, frankness in writing "Planting Bamboos".

That "Heavenly Palace" is used to refer to XJTU not only lies in its charming culture, but in a group of extraordinary masters, including Zhong Zhaolin, father of electrical motors in China, Chen Daxie, a pioneer in thermal engineering, Chen Xuejun, an energy and power engineering scientist, and Qu Liangsheng, the founder of mechanical fault diagnosis.

This land has witnessed great changes in history, from the powerful Western Han dynasty to the prosperous Tang dynasty, and to the modern city at present. This century-old university brought together the pioneers who moved from Shanghai to western China for national construction, the young talents from all over the world, and the dream pursuers of the development of western China.

Let's start the journey together to explore the fascinating stories, which happened at this site, to meet those whose names ring a bell.

Let's go! Wish this trip to *Xian Jiaoda*, the Heavenly Palace memorable and worthwhile.

# 请收下我的名片

姓名：西安交通大学

出生地：上海

出生年份：1896年

生日：4月8日

曾用名：南洋公学 南洋大学堂
交通部上海工业专门学校 交通大学
交通大学（西安部分）等

主营业务：人才培养、科学研究、社会服务
文化传承创新、国际交流与合作

身份标签：① 开拓者：我国最早创办的高
等学府之一、中国高等工程教育的重要源头
之一；② 实干家：1956年响应党中央、国务院
的号召内迁西安，成为西部大开发的先行者；
③ 国家队：首批"211""985""双一流"建设
学校，C9联盟发起人之一，"珠峰计划"
"卓越工程师教育培养计划"首批高校。

地址：陕西西安—兴庆校区、雁塔校区、曲江校区
陕西西咸新区—中国西部科技创新港

座右铭（校训）：精勤求学 敦笃励志
果毅力行 忠恕任事

人生格言（办学定位）：扎根西部 服务国家
世界一流

个人主页：www.xjtu.edu.cn

# Name Card of XJTU

Name: Xi'an Jiaotong University
Birthplace: Shanghai
Date of Birth: April 8, 1896

Former Names: Nanyang Public College, Grand Nanyang University, Shanghai Special Industrial School of the Ministry of Transportation, Jiaotong University, Jiaotong University (Xi'an Campus), etc.

Major Business: talents cultivation, scientific research, social services, cultural inheritance and innovation, international exchanges and cooperation

Identity Labels:
1. Pioneer: One of the earliest institutions of higher education in China, the origin of advanced engineering education;
2. Practitioner: Pioneer of western China development in response to the state call of relocating to Xi'an in 1956;
3. National team member: One of the first batch of China's "Project 211" and "Project 985" universities; on the list of "Double First-Class" universities (World-class Universities and World-class Disciplines); one of the founders of Universities of China's C9 League; one of the first batch of universities of "Mount Qomolangma Plan" (a gifted college program to cultivate excellent innovative talents) and "The Education and Training Program of Excellent Engineers (ETPEE)."

Address:   Xingqing Campus, Yanta Campus, Qujiang Campus, Xi'an, Shaanxi Province;
Western China Science and Technology Innovation Harbour (iHarbour), Xixian New Area, Shaanxi Province

Motto:     Diligence, Ambition, Decisiveness, Loyalty

Orientation:  Taking root in western China, providing services to the nation, and striving to be a world-class university

Website:   www.xjtu.edu.cn

# 目录 Contents

3

# 穿越路线一
## 西迁精神路线

# Route 1
Route of Westward Relocation Spirit

# 穿越须知

# What You Must Know
# Before Starting the Journey

### 何为交大西迁？
1956 年，根据党中央、国务院决定，在上海生长了 60 年的交通大学内迁西安，从此便永久扎根在祖国的西部。

### 为什么要西迁？
从短期看，是为了支援西北工业基地建设和国防事业发展；从长远看，是为了新中国的教育布局、工业布局，拉动、推动西部的发展，由党中央、国务院作出的战略部署。

### 什么是西迁精神？
核心：爱国主义
精髓：听党指挥跟党走

### 西迁产生了何种影响？
交大西迁改变了整个中国西部高等教育的格局。西安交通大学通过自身的发展壮大，引领和带动整个西部地区高等教育乃至全国教育的蓬勃发展，形成了"一马当先，万马奔腾"的大好局面。

### 为何说"五所交大是一家"？
交通大学 1921 年定名，时分京校、沪校和唐校三地。几经分合，历经战火，最终在西安、成都、北京、上海及台湾新竹发出新枝。西安交通大学、上海交通大学、北京交通大学、西南交通大学、台湾新竹交通大学五所交大同根同源，交融互通。

### Q: What does westward relocation of Jiaotong University mean?
A: Westward relocation refers to the national strategical decision that the main body of Jiaotong University be moved from the affluent metropolis of Shanghai in east China to the resource-strapped northwestern city of Xi'an in 1956 to serve the national strategy of industrial development in the western region.

### Q: Why was it decided to move Jiaotong University to Xi'an?
A: In the short term, westward relocation was to support the construction of northwest industrial bases and the development of national defense. While in the long term, it was a visionary strategic layout adjustment of education and industry made by China's central government in order to promote the development of the western regions in China.

### Q: What is essence of "Westward Relocation Spirit"?
Core: Patriotism
Essence: Follow the Communist Party of China's order

### Q: What is the influence of westward relocation?
A: The westward relocation of Jiaotong University has changed the pattern of higher education in western China. Through its own development, XJTU has played a leading role in promoting the vigorous development of higher education and education as a whole in the western regions of China.

### Q: Why is it said that five Jiaotong universities are of one family?
A: The name of Jiaotong University was designated in 1921, with three branch universities in Beijing, Shanghai and Tangshan respectively. After several separations and merges, wars and changes, Jiaotong University eventually developed into five independent universities in Xi'an, Chengdu, Beijing, Shanghai and Hsinchu, respectively. These five Jiaotong universities are of the same origin and communicate with each other to promote common development.

# 1.
# 北门

位置 | 校园最北部

## The North Gate

The North End of the Campus

忆来惟把旧书看，几时携手入长安？

交通大学初来西安时，此地尚为一片麦田，麦浪翻滚，即将迎来收获。而此处便是西迁时学校正门。历尽春秋，校门也曾几度更迭。

西迁正逢第一个五年计划大规模基础建设时期，国家倡导增产节约，为了迎接迁校师生和迁校后第一个国庆节，学校仿照 1955 年国庆节时上海华山路老交大校门口"国庆牌坊"结构，以梯形竹木临时赶制成落户西安的首个校门。这两座建筑都出自学校总务处老木工张师傅之手，造型古朴典雅、端庄凝重。一批又一批乘坐"交大支援大西北"专列的交大学子便从这里踏入校园，开启逐梦旅程。1960 年前后，新校门开建，为三拱联结型木结构，成为西迁师生毕业留念的必去"打卡地"之一，深深留在了大家的记忆中。其后多年，因风雨侵蚀几经修葺，后重建校门，以砼立柱替代，柱顶饰以"三面红旗"，象征特有的政治含义。现在所用校门乃1992 年启用，正面上方为圆弧形，上书毛体"交通大学"校名及英文翻译；两侧为浮雕群"慧识界"，采用基础科学、技术科学和应用科学之典型符号，代表交大"理工管"结合之办学特色，体现"集大成，得智慧"的育人理念，气势雄伟，意蕴弘深。

When Jiaotong University was just relocated to Xi'an, the selected site for campus was a field of ripe wheat, about to harvest. And right here was the main entrance of the university in spite of its undergoing several changes.

The university relocation coincided with large-scale infrastructure construction during the first Five-Year Plan in China, when production increase and resource saving were advocated. Therefore, the first school gate, modeled on the structure of "the National Day Memorial Archway" in front of the entrance of the former Jiaotong University at Huashan Road, Shanghai during the National Day in 1955, was temporarily made with trapezoid bamboo, to celebrate the first National Day after the relocation and welcome teachers and students to the new university. Both gates, designed and made by an experienced carpenter working in the General Office of the university, were simple, classic and elegant, through which generations of students entered the university and started their dreams-seeking journey. Around 1960, a new three-arch wooden school gate was built and became one of the must-visit places for graduation pictures. The gate has been undergoing renovations over the past three decades. The north gate at present was put into use in 1992. It is of a circular arch frame, with "JIAOTONG UNIVERSITY" in both Chinese and English inscribed on it. The gate is flanked with magnificent anaglyphs on two sides— "World of Wisdom". The anaglyphs combine the classic symbols of basic science, experimental science and applied science, which represents the characteristics of the integration of "science, engineering and management" of Jiaotong University, and embodies the profound educational concept of "integrating great achievements and obtaining wisdom".

上：1896年南洋公学（交通大学前身）创立初期校门　中：1935-1956年交通大学校门　下：1956-1958年交通大学西安部分校门　**003**

上：1958-1968年交通大学西安部分、西安交通大学校门　中：1959年电气绝缘、电缆技术专业毕业班欢度国庆留影　下：1962年校历

上：1968 年校门　中：1984 年校门　下：1992 年新校门（北门）建成

# 2.
# 饮水思源碑
## 位置 | 中心楼北

# Yinshui Siyuan Stele
North to the Central Teaching Buildings

落其实者思其树，饮其流者怀其源。

饮水思源池初建于 1933 年，是 1930 届学生在毕业之际为感念母校培育之恩所建。喷水池中央矗立有"饮水思源"碑石，碑石上方是立体的交大校徽。校徽为圆形，中有一铁砧，向左斜靠着一把铁锤，右有五节链条相垂，上面摆放着中西书籍数本，取寓工于读、学重中西之意。铁砧底座上刻有创校年代 1896 年，谓奠基之始。环盾作齿轮形，略似电机中转核之横截面，表示学科，也可意为交通事业。外圈刻有学校中英文校名。西迁后，学校在西安新校园复建"饮水思源"碑，时名"饮水思源留念塔"。

1981 年 4 月，在即将迎来 83 周年校庆的时候，学校决定重建饮水思源碑，同年 10 月饮水思源碑便在北门内、中心楼前水池中正式落成。碑石"饮水思源"四字由中国书法家协会第一任主席舒同题写。碑石背面汉白玉碑文记载了交通大学于 1896 年创办以来的沿革与变迁。塔身、塔基用红黑色磨光花岗岩石和汉白玉石料组合而成。顶端的老交大校徽、齿轮、铁砧、丁字尺、书本造型由青铜铸造而成。塔基周围与水池接触部分是墨玉磨光花岗岩石，并设计了花坛，可种植花卉，塔基四角有四只莲花式喷水头。

2020 年，一座崭新的饮水思源碑矗立在西安交通大学中国西部科技创新港校园内。是的，不论是在西安交大，还是在上海交大、北京交大、西南交大以及新竹交大的校园内，你都能遇见"饮水思源"的身影。这四个字早已融入交大血脉，化为交大文化传统的内在基因，薪火相传。

When eating fruits, one should think of where they are from and when drinking water, think of its source.

Yinshui Siyuan Fountain was built by Class of 1930 graduates for the gratitude of *alma mater* in 1933. In the center of the fountain stands the Yinshui Siyuan (when you drinking the water, think of its source) Stele, on top of which is the three-dimensional emblem of Jiaotong University. The school emblem is round-shaped with an anvil in the middle which leans against a hammer to the left and has five chains hanging down to the right. On the anvil are several Chinese and Western books, which means well-versed in the learning of both Chinese and Western technologies. The anvil base is engraved with 1896, the founding year of Jiaotong University. The ring shield is shaped like a gear, somewhat like the cross section of an electric motor armature, representing a discipline or engineering and transportation. The outer of the ring is engraved with the name of the school in Chinese and English. After westward relocation, the stele was rebuilt on Xi'an campus, known as "Yinshui Siyuan Memorial Tower" then.

In April, 1981, before the 83rd anniversary of the university, XJTU decided to rebuild the stele. In October of the same year, the construction of new stele was completed in the pond in front of the Central Teaching Buildings at the north gate. The inscription of "Yinshui Siyuan" was written by Shu Tong, the first chairman of Chinese Calligraphers Association. On the back of the stele is the white marble inscription of the evolution and changes of Jiaotong University since it was founded in 1896. The body and base of the stele are made of red and black polished granite and white marble. On the top of the stele are the university emblem, gear, anvil, T-square and books, made of bronze. Around the granite stele base is a flower bed, and four lotus-style sprinklers are set at the four corners of the base.

In 2020, a brandnew Yinshui Siyuan Stele stood at the Campus gate of Western China Science and Technology Innovation Harbour of XJTU (iHarbour). Actually, not only in XJTU, you can also find Yinshui Siyuan Stele on campuses in other four Jiaotong universities in Shanghai, Beijing, Chengdu, and Hsinchu. These four characters "Yinshui Siyuan" have become the gene of Jiaotong University's cultural tradition, passed on from generation to generation.

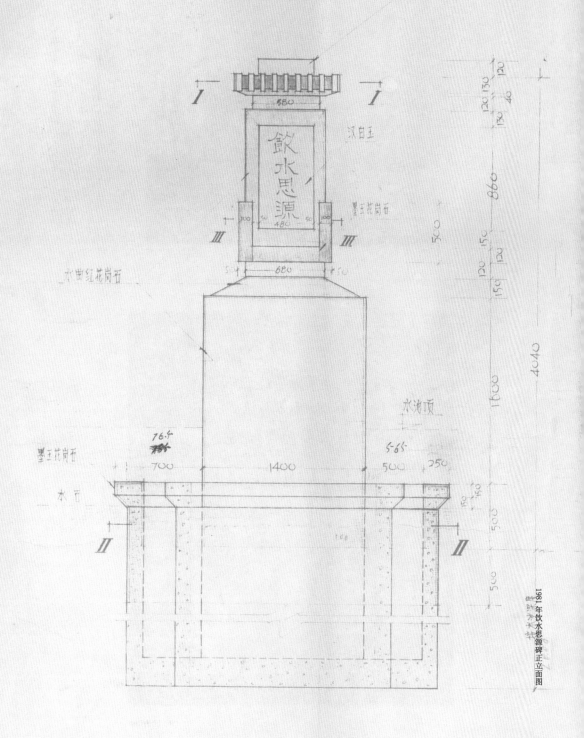

上部为铸铜齿轮

汉白玉

饮水思源
480

墨玉花岗石

水曲红花岗石

墨玉花岗石

水面

水池顶

正立面图

1981 年饮水思源碑正立面图

一组关于饮水思源碑的记忆

# 3.
# 雪松
位置 │ 中心楼、东西花园等

## Cedars
Around the Central Teaching Buildings,
Inside the East & West Gardens

大雪压青松，青松挺且直。

走在校园里，到处都能看到一棵棵巨大挺拔的雪松，撑开枝伞，如守卫一般守护校园。其中一种名为喜马拉雅雪松，西迁时采购种植于此。据说当时市场价超过数百元人民币，而那时一级教授钟兆琳的工资也仅有 300 元左右。如果让交大人投票选出最富交大气质的植物代表，雪松一定会名列三甲，它代表了交大的精气神。

刚来西安时，电灯不明，电话不灵，马路无风三尺土，有雨满街泥。西迁伊始，交大人就想在西北建设一所绿树成荫、环境优美的江南范大学。从开始基建时就由西北农学院设计绿化，并从上海、苏州、扬州、杭州、宁波等地采购各种花木数十万株，如雪松、罗汉松、白玉兰、红枫、法国梧桐、紫丁香、女贞、木槿、大叶黄杨球等不计其数。后来总务处专门开辟苗圃，培育山茶、杜鹃、龙柏等树苗。师生齐动员，共同建设花园式校园，是当时西迁师生的必修劳动课。学校在二十世纪八九十年代多次受到中央绿化委员会表彰。

Walking on the campus, you can see tall cedars everywhere with huge spreading branches and leaves. A cedar called Himalayan cedar was planted here after westward relocation. It is said that it was worth 1000 yuan RMB in the market, much more than the monthly salary 300 yuan RMB Grade 1 Professor Zhong Zhaolin was paid. If there were a vote for a plant representative of the temperament and spirit of XJTU, cedar would definitely rank among the top three.

Since the relocation, XJTU planned to build a beautiful campus full of trees and plants in northwest region like those in southern China. So the campus greening was designed by National Northwest Junior College of Agriculture and Forestry since the start of construction, and hundreds of thousands of valuable plants and trees were purchased from Shanghai, Suzhou, Yangzhou, Hangzhou and Ningbo, including cedar, podocarpus macrophyllus, magnolia, red maple, plane tree, lilac and so on. Later, the General Office cultivated a piece of land for camellia, azalea, dragon juniper and other saplings planting, mobilizing all teachers and students in building a beautiful campus.

上：1958 年校景一角　下：1964 年校园绿化劳动

# 4.
# 中心楼
位置 | 腾飞塔广场北

## The Central Teaching Buildings
North to the Tengfei Tower Square

高楼一何峻,迢迢峻而安。

中心楼号称"西迁第一楼",位于教学区中央,始建于 1956 年,总建筑面积 3 万余平方米。

1955 年 4 月,交大接到中央关于西迁的指令,次年秋季便要在西安开展正常教学活动。由于设计关系,中心楼开工建设较晚,加之西安市基本建设规模庞大,基建材料供应紧张,时建时停,工期因此更加紧迫。西安市委积极组织人力物力,配合学校基建各工组,齐心协力,日夜奋战,终于赶在开学前一个月基本完工,保障了顺利开学。交大西迁历时三年,没有因为迁校而推迟一天开学,没有少设一门课,耽误一次教学实验,不少人都感叹这是一个奇迹。中心楼的顺利完工就是其中的一个缩影,堪称陕西省、西安市与学校精诚团结、共克艰难的典范。

时代的象征在建筑上得以凝固。中心楼采用了简洁的"中苏风格"——青砖立面、"人字形"红瓦坡顶、中轴对称一体两翼合围式布局。中心楼是西迁基建的核心工程,承担着一二年级基础课教学和电工原理、材料加工等基础课程实验功能。2014 年,中心楼被列入第六批陕西省级文物保护建筑。

The Central Teaching Buildings, also known as the "First Buildings after Relocation to Xi'an", built in 1956, are located at the center of the teaching area with a total construction area of over 30,000 square meters. The construction of the building, with coordination and concerted efforts from varied departments of Xi'an city and the university, was basically completed one month earlier before September, 1956 , to make sure that the new semester would start as scheduled. It took three years for Jiaotong University to finish relocation to Xi'an, yet there was no single class delayed, neither one single teaching experiment canceled in the process, which was praised as a miracle.

The Central Teaching Buildings adopted the simple Chinese-and-Soviet style with dark gray brick facade, the "herringbone" red tile slope top and a symmetrical layout with a central axis and two wings enclosed. A core project in the process of relocation, the Central Buildings played a key role for first-and-second-year students to have basic courses and basic experimental courses including "Principles of Electric Engineering" and "Material Processing". In 2014, this building was listed as the sixth batch of Shaanxi provincial-level cultural relic under protection.

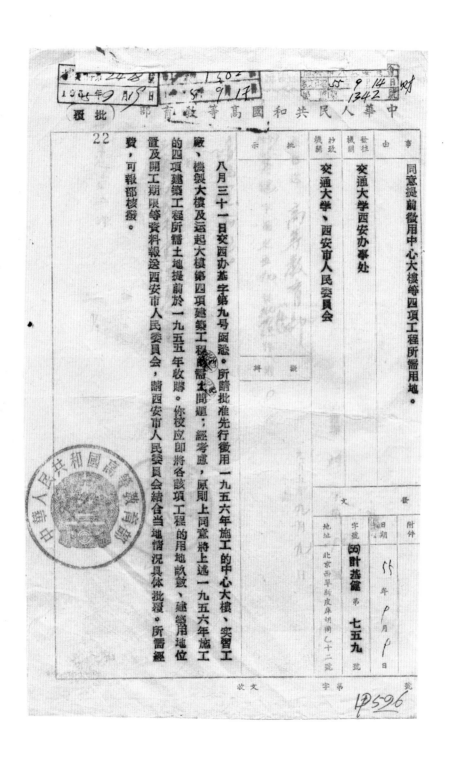

中華人民共和國高等教育部

22

事由　同意提前徵用中心大樓等四項工程所需用地。

發往機關　交通大學西安辦事處

抄致機關　交通大學、西安市人民委員會

批示

八月三十一日交西辦基字第九号函悉。所請批准先行徵用一九五六年施工的中心大樓、實習工廠、機製大樓及運起大樓第四項建築工程的需土問題：經考慮，原則上同意將上述一九五六年施工的四項建築工程所需土地提前於一九五五年收購。你校應即將各該項工程的用地欸數、建築用地位置及開工期限等資料報送西安市人民委員會，請西安市人民委員會結合当地情況具体批覆。所需經費，可報部核撥。

簽發日期　55年8月8日

字號　(55)計基健第七五九號

地址　北京西單舊库胡同乙二二号

附件

收文字第　IP596號

　　　上：1959 年中心楼全景　　下 1：西迁教授张寰镜指导学生制图　　下 2：改革开放初学校中心楼门前

下3：1980年电化教学　下4：1983年金相显微镜实验室

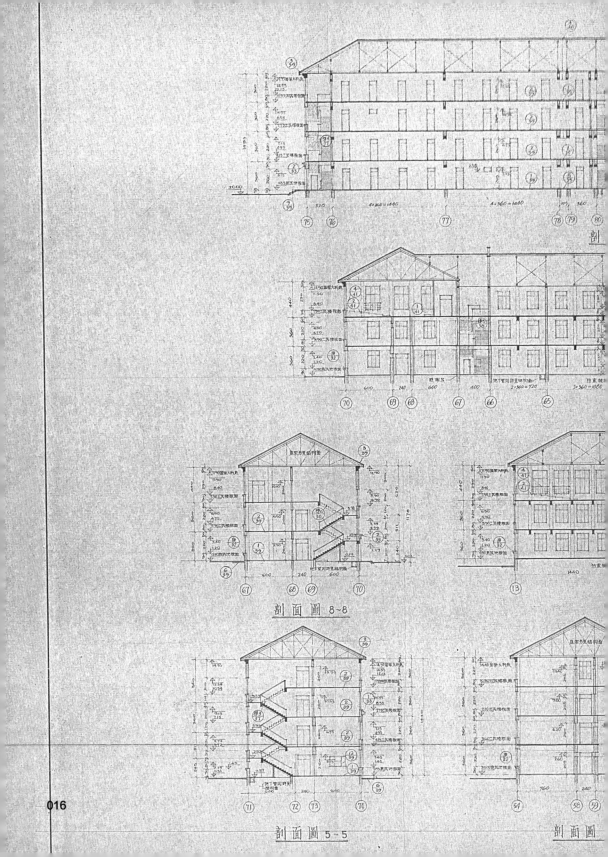

剖面圖 8~8

剖面圖 5~5

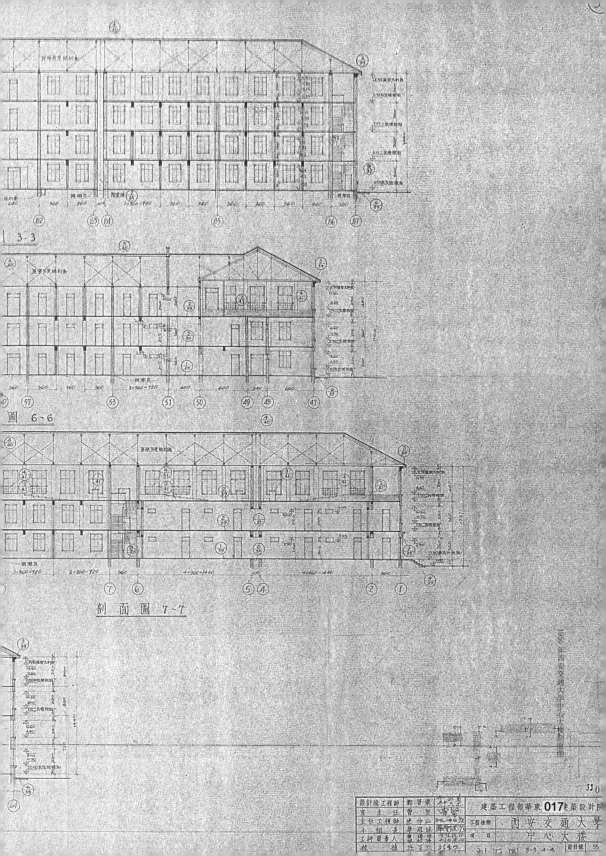

剖面图 7~7

1956年西安交通大学中心大楼剖面图

建筑工程部华东 017 建筑设计院

工程總册 西安交通大學
項目 中心大楼

# 5.
# 杰出校友画廊
位置 | 中心楼连廊

# Hall of Fame for Alumni
At the Vestibule of the Central Teaching Buildings

须知吾人欲成学问，当为第一等学问；欲成事业，当为第一等事业；欲成人才，当为第一等人才。而欲成第一等学问、事业、人才，必先砥砺第一等品行。

这是交通大学工程教育奠基者唐文治校长提出的人才培养观。在国学大师唐文治的带领下，交大在高等工程教育方面大胆开拓、敢为人先，成为当时独一无二的新型工业大学。唐文治老校长提出的"实心实力求实学、实心实力务实业"的办学思想，"造就中国之奇才异能，冀与欧美各国颉颃争胜"的办学目标，以及"第一等"的人才培养观，如今读来，依然振聋发聩。

风云两甲子，群星耀苍穹。从交大校园里走出去的，有辛亥革命时期叱咤风云之人物、护国元勋蔡锷，喋血辛亥的爱国地理学家、共产党人李大钊的民主思想启蒙者白毓崑，人民科学家、"两弹一星"元勋钱学森，"中国核潜艇之父"黄旭华，国际拓扑学大师吴文俊，中国水电泰斗张光斗，"中国稀土之父"徐光宪……西迁65年，西安交大累计培养了28万名毕业生，其中两院院士就有近50位。

Tang Wenzhi, former president and founder of engineering education of Jiaotong University (Shanghai), was a great educator and master of Chinese culture. He put forward the goal of building the first class university in China with emphasis on practice and application of knowledge. He made every effort to establish engineering education with foresight and sagacity, advocating to lay equal stress on engineering education and liberal arts and dedicating to the integration of traditional cultural resources with modern scientific education, which made Jiaotong University a leading university of engineering with unique characteristics.

Cai E, a military general, a Chinese revolutionary leader in Xinhai Revolution of 1911; Bai Yukun, the patriotic geographer and the enlightener of democratic thought; Qian Xuesen (Tsien Hsue-shen), an eminent scientist and the founding father of Chinese rocketry and spaceflight; Huang Xuhua, the chief designer of China's first generation of nuclear submarines; Wu Wenjun, the master of topology in the world; Zhang Guangdou, the leading expert in China's water conservancy and hydropower engineering; Xu Guangxian, the father of China's rare earth.

During the 65 years of westward relocation, XJTU has graduated 280,000 graduates, nearly 50 of whom became academicians of Chinese Academy of Sciences and/or Chinese Academy of Engineering.

1959 年钱学森学长回母校

# 國立交通大學

第 二五 學年度第 一 學期　
科　學院　學院　學　門系　　年級 一 年級　系　科
中華民國廿五年 一 月　考試

| 號數 | 姓名 | 國文 (2) | 英文 (4) | 體育 (1) | 物理講授 (4) | 物理試驗 (2) | 化學講授 (4) | 化學試驗 (2) | 數學通識 (2) | 國形幾何 (4) | 軍事訓練 (0) | 學分總積 | 平均成績 |
|---|---|---|---|---|---|---|---|---|---|---|---|---|---|
| 350 | 嚴廣駿 | 131.28 / 65.64 | 68 40 / 50 | 90 | 282 / 70.5 | 140 / 70 | 240 / 60.0 | 140 / 70 | 88 / 57.8 | 81 | 73 | 1664.28 | 69.35 |
| 420 | 錢鴻業 | 138.28 / 69.14 | 304 / 76 | 85 | 320.4 / 80.1 | 150 / 75 | 344 / 86.0 | 176 / 88 | 304.8 / 76.2 | 86 | 85 | 1928.48 | 79.52 |
| 422 | 裘家康 | 142.32 / 71.16 | 244 / 61 | 90 | 242 / 60.5 | 156 / 78 | 253.6 / 63.4 | 140 / 70 | 244.4 / 61.1 | 64 | 74.5 | 1515.42 | 63.19 |
| 423 | 鄭諫華 | 147.28 / 73.64 | 276 / 69 | 85 | 283.2 / 70.8 | 160 / 80 | 307.2 / 76.8 | 158 / 79 | 246 / 61.5 | 82 | 75 | 1744.40 | 72.72 |
| 424 | 李水良 | 133.72 / 66.86 | 276 / 69 | 85 | 330.4 / 82.6 | 160 / 80 | 320 / 80.0 | 170 / 85 | 720.0 / 72.0 | 78 | 75 | 1841.12 | 76.71 |
| 425 | 陸孚同 | 157.90 / 78.95 | 264 / 66 | 80 | 272.4 / 68.1 | 160 / 80 | 267.2 / 66.8 | 150 / 75 | 280 / 58.2 | 75 | 84.5 | 1674.20 | 69.75 |
| 426 | 羅國灝 | 151.4 / 75.70 | 252 / 63 | 80 | 284 / 71.0 | 160 / 80 | 255.2 / 63.8 | 148 / 74 | 294.4 / 73.6 | 74 | 76 | 1699.00 | 70.79 |
| 427 | 趙孟養 | 160.44 / 80.22 | 340 / 85 | 80 | 240.4 / 60.1 | 160 / 80 | 287.2 / 71.8 | 160 / 80 | 300.8 / 75.2 | 78 | 74 | 1952.04 | 75.02 |
| 428 | 李壽萩 | 151.6 | | | | | 391.8 | | | | | | |

| | | | | | | | | | | | |
|---|---|---|---|---|---|---|---|---|---|---|---|
| 73.73 | | | | | | | | | | | 劉復祥 | +30 |
| 73.40 | 1607.01 | | 82 | | | | | 80 | 147.76 7388 | | |
| | 1205.72 | | | | | | | 88 | 296 74 | 洪維懷 | +31 |
| 67.49 | 1855616 | | | | | | | 84 | 1744 6270 | 劉維藩 | +32 |
| 71.25 | 1718.00 | 69 | 76 | | | | | 80 | 12856 6428 | 陸 正 | |
| 68.81 | 1651.36 | 73 | 66 | | | | | 87 | 1268 67 | 林衍光 | +33 |
| 82.71 | 1793.12 | | 86 | | | | | 80 | 158.12 7906 | 楊祖胎 | +34 |
| 67.36 | 1616.72 | 755 | 62 | | | | | 78 | 1379.2 6896 | 黃禮鎮 | +35 |
| 66.97 | 1572.02 | 645 | 68 | | | | | 80 | 130.36 6718 | 徐 質 | +36 |
| 73.68 | 1768.74 | 715 | 88 | | | | | 85 | 13144 6572 | 林致德 | +37 |
| 77.53 | 1852.64 | 745 | 76 | | | | | 92 | 13084 6542 | 陳繼能 | +38 |
| 68.39 | 1641.36 | 77 | 60 | | | | | 85 | 13216 6608 | 徐賢良 | +39 |
| 71.31 | 1711.44 | 765 | 86 | | | | | 85 | 12584 6542 | 吳文俊 | +40 |
| 72.71 | 1745.12 | 715 | 75 | | | | | 87 | 1427.2 71.36 | 陸繼芳 | +41 |
| 78.09 | 2030.40 | 755 | 82 | | | | | 82 | 1358.0 6770 | 沙成玉 | +42 |
| 72.2 | 1733.96 | 805 | 65 | | | | | 85 | 1421.6 7108 | 郁青田 | +43 |
| 65.35 | 1572.36 | 825 | 82 | | | | | 86 | 131168 6584 | 童紹榮 | +44 |
| 66.62 | 1599.00 | 73 | 65 | | | | | 85 | 1258.0 6290 | 詹倧恒 | +45 |
| 71.0 | 1706.24 | 61 | 70 | | | | | 82 | 131.44 6572 | 祝宗壽 | +46 |
| 67.9 | 1702.8 | 70 | 60 | | | | | 82 | 141.68 7084 | 王有輝 | +47 |
| | | 615 | 86 | | | | | 85 | 1297.2 6476 | 讚 生 | +51 |
| | | | | | | | | | | 試 | 試 |
| 73.93 | 1647.4 | 77 | 70 | | | | | 80 | 141.28 7064 | 沈立銘 | +48 |
| | | 825 | 78 | | | | | 87 | 139.6 6980 | 王正術 | +50 |
| | | | | | | | | | | 科 士 | 準 |

# 6.
# 校风墙

位置 | 中二楼北

## The Presenting Wall of Campus Climate
North to Central Teaching Building 2

承传讲学风，刻厉饬伦纪。

校风是学校的风气，它看似不可捉摸却又无处不在，就像呼吸和吐纳的空气一般，浸淫着、影响着生活在校园里的师生。

西安交大的校风——爱国爱校、追求真理、勤奋踏实、艰苦朴素，由杰出校友陆定一1985年所题。他认为，这是从交大同志们几十年辛勤劳动中总结而来，是教育界、工业界可以借鉴的。

陆定一，中国共产党宣传思想战线的卓越领导人。1926年毕业于交通部南洋大学电机工程科，是交通大学第一个党团支部的重要成员。1927年任共青团中央宣传部部长。中华人民共和国成立后，历任中共中央宣传部部长、国务院副总理、中央书记处书记、文化部部长（兼）、全国政协副主席。交大浓厚的革命文化氛围和育人传统，给他极深影响，他说，"我曾在交大读书八年，我的共产主义世界观是在那里形成的"。当时学校"重视招生质量，坚持择优录取；重视基础理论教学和基本技能训练；对学生严格要求、严格考核，强调理论结合实际，学以致用"。20世纪80年代，陆定一担任西安交大校务委员会主任时，将交大的办学传统和办学特色概括为：起点高、基础厚、要求严、重实践。

The campus climate is to the teachers and students what air is to people.

The campus climate of Xi'an Jiaotong University—"patriotism, love for university, pursuit of truth, diligence, hard work and plain living", was written by Lu Dingyi in 1985, an outstanding alumnus of XJTU. He believed that this was summarized based on decades of efforts of people in Jiaotong University, and can be used for reference in educational and industrial circles.

Lu Dingyi, one of the top officials of the CPC, graduated from the Electrical Engineering Department of Nanyang Public College of Ministry of Transportation in 1926. After the founding of the PRC, he served as Vice Premier of the State Council, Secretary of the CPC Secretariat, Minister of Culture and Vice Chairman of the CPPCC. Deeply influenced by and greatly benefited from the culture and educational tradition of Jiaotong University, Lu Dingyi, the then Director of XJTU School Council in 1980s, generalized the traditions and school running characteristics as a higher starting point, a more solid foundation, more strict requirements, and more emphasis on practice.

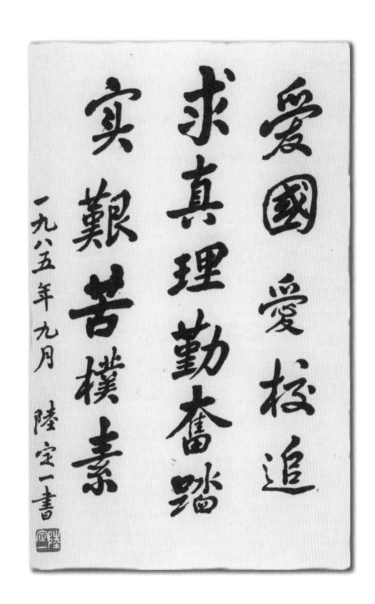

爱国爱校

求真理勤奋踏

实艰苦朴素

一九八五年九月　陆定一书

1985 年 9 月陆定一学长题写的西安交通大学校风

# 7.
# 西迁广场
位置 ｜ 西二楼西

# The Westward Relocation Square
West to West Building 2

秦中海上西征路，万里驱车亦壮哉。

"火车在飞奔，车轮在歌唱。装载着木材和食粮，运来了地下的矿藏。多装快跑，快跑多装。把原料送到工厂，把机器带给农庄……"

1956年8月10日，在"交大支援大西北专列"上，《我们要和时间赛跑》的歌声飘荡在列车的车厢中。上千名交大师生从上海启程，向西安进发。这是注定要被载入史册的时刻，欢送的队伍十分壮观，大轿车从学校出发，敲锣打鼓经过南京路，一直到上海北站。车站人山人海，送行的人几乎与登车队伍相等。上海其他高校师生闻讯赶来送别，复旦大学、同济大学等校师生将校徽取下来别在校旗上为交大师生留作纪念。这一幕就烙印在了西迁广场的浮雕上。

西迁广场背景墙上题刻"交通大学西行之履"，上塑西迁精神、交通大学西迁述略、迁校各类机构设置沿革等文字内容。"西行之履"主题雕塑上的脚印，象征西行的脚步坚定有力，疾速迈进。广场内另设迁校浮雕六幅，内容分别为：上海外滩、交通大学上海徐家汇校区鸟瞰、迁校时期西安城区鸟瞰、西安校区鸟瞰、上海火车站迁校送别场景、西安迎接迁校师生场景。安静的广场诉说着无言的故事，西迁的传奇正是由这一个个平凡却又伟大的人物书写的。

广场旁乃创校纪念校门——南洋公学牌楼，根据南洋公学老校门建筑风格建成。校门牌匾分别刻"南洋公学"与"交通大学"，根据1906年前南洋公学校门牌匾原迹，以及1928—1935年交通大学校门牌匾原迹进行调整、复制而成。

August 10, 1956 will surely be recorded in the history of XJTU, when thousands of academic faculty and students of Jiaotong University left Shanghai for Xi'an by the Special Train for Northwest China. The scene that large crowds of people, including teachers and students from other universities in Shanghai, coming to bid farewell was carved on the reliefs of the Westward Relocation Square.

On the background wall of the Westward Relocation Square was inscribed the "Jiaotong University's Journey to the West", with 16 characters of the content of westward relocation spirit, a brief introduction of the westward relocation, and the evolution of various institutions set up for the relocation. The footprints on the sculpture of "Journey to the West" symbolize the firm pace of marching westward. In the square, there are six relief works, which are about: The Bund (Shanghai), an aerial view of Xujiahui Campus of Jiaotong University in Shanghai, an aerial view of Xi'an city during the relocation of the university, an aerial view of Xi'an Campus, the scene of bidding farewell in Shanghai Railway Station, and the scene of welcoming teachers and students from Shanghai to Xi'an. The square witnessed what happened on the new campus, and the legend of westward relocation was written by these ordinary but great people.

Beside the square is a commemorate gate for establishment of the university—Archway of Nanyang Public College, which was built according to the architectural style of the school gate of Nanyang Public College. The school gate plaques were engraved with "Nanyang Public College" and "Jiaotong University" respectively, which were duplicated and adjusted based on the original plaque of Nanyang Public College before 1906 and the original gate plaque of Jiaotong University from 1928 to 1935.

### 西安新校服务商場開幕

我校西安新校舍服务商場已竣工，十一月一日已有百货商場（包括布匹、糖果、文具、日用品、油醬、土產等）及服裝部、洗染部、新華書店、粮站等遷入營業，还有人民銀行、邮局、皮鞋部、理髮部等亦將于本月內先后遷入營業。

### 由上海來校的一批商店開始營業

为支援我校西遷，上海市商業局抽調一批商店随我校來西安營業，其中有的已開始營業。

属上海天和煤球厂來校經過近月的籌备，克服了机件不全、厂房缺乏的困难，已于九月上旬開工生產。除供应本校外，还满足了西安動力学院、西安航空学院教工的需要。

洗染商店工人員14人最近也已分别在員工宿舍及学生宿舍營業，設有洗燙、縫補、雨衣上膠等，不久还將增加染色部，价格比上海及西安一般商店便宜。

理髮部工作人員26人也已分别在員工宿舍及学生宿舍營業，不久將备有男女电燙、水燙、火燙、蒸髮等設備。

此外，成衣部、皮鞋部也正在籌建中，有三十余年工齡的時裝、西部裁剪工人二名及制鞋業職工十余人不久即从上海來校。

賦利房屋即將完工，完工后这批商部即遷入營業。（陸明）

### 又有兩个商店在西安新校址開始營業

我校西安部分的成衣部已經開始營業。成衣部分中式、西裝、時裝、服裝四个部分，共有九个裁剪縫紉工人，其中有三人已有二十年以上的工齡。十五日是他們開始營業的第一天，这天他們接受了150件衣裳的剪裁工作，營業额達三百元。校內成衣部的价格比上海、西安的一般服裝店要便宜。

西安部分皮鞋部也已經開始營業。从上海來的三名工人，將为交大的師生員工們制作皮鞋布鞋和修理皮鞋等。

一九五五年十一月十日　　　大交　　　第三版

#### 西安新校在建設中

我校西安新校址已開工興建

---

上：1956年交通大学西迁师生的乘车证　下：西迁时的部分校报报道

025

左上：西迁初期足球队在训练　　中上：1959 年学生跳伞练习　　中：西迁初期举重队在训练　　左下：1956 年教工摩托车训练班培训
中下：英姿飒爽的摩托车队　　右上：1959 年校运会入场仪式　　右中：1959 年女篮合影　　右下：1959 年校运会跳高比赛

　左上：1959年伊拉克学生代表团来访　右上：1959年图书馆给书籍编目上架　左下：1957年教学研讨会　右下：1958年材料力学教研组探索电化教学

上：1959年，钟兆琳教授指导青年教师　中：1958年几何工程画模型展览　下：陆庆乐教授在上课

西安交通大学 120 周年校庆贾濯非教授创作的西迁纪念广场主体浮雕手稿

# 8.
# 樱花道
位置 | 钱学森图书馆东、西侧

## Cherry Tree Roads
East and west sides to Qian Xuesen Library

樱花红陌上，杨柳绿池边。

每年3月底4月初，旖旎春光里的烂漫樱花成为交大最美丽的景色，也是古城西安"藏在深闺"中的胜景。届时两条樱花道游人如织，摩肩接踵，风吹落英，樱羽弥漫，好不热闹。

兴庆校区的樱花树共有340余棵，主要分布于校园樱花东道、樱花西道以及图书馆两侧，颜色以淡粉色居多，亦有少数白色、黄色和绿色树种。樱花树在20世纪60年代前后栽植，因树龄老化等原因曾两度更新——在90周年校庆和110周年校庆前夕，学校从浙江两次购回樱花树苗栽植。两条樱花道一直保持着茂盛的青春气息，如树下走过的学子一般永远朝气蓬勃。

The end of March and the beginning of April each year is cherry-blossom season, presenting the most beautiful scenery of XJTU and attracting many visitors to gather here.

There are more than 340 cherry trees on campus, mainly distributed on the East and West Cherry Tree Roads and both sides of the library. Most of the trees are light pink, and a few are white, yellow and green. The cherry trees were planted around 1960s, and were replanted before the 90th and 110th anniversary. So the two Cherry Tree Roads have always maintained lush, like the students walking under the trees, always vigorous.

上：20世纪70年代樱花道　下：20世纪90年代樱花道

# 9.
# 梧桐道
位置 | 中心楼东西侧

# East and West Plane Tree Roads
East & West Sides to the Central Teaching Buildings

维梧有凤，鸣于朝阳。

当秋天的调色盘打翻，"仙交大"就会呈现出别样的缤纷韵致，这时，交大人会专门在梧桐道留出一条落叶路，等待金色的梧桐落叶飘满，诗意的梧桐道和走在梧桐道上的师生便成了校园的一道风景，这便是"金色梧桐 诗意交大"的由来。

梧桐代表交大的气质。这些法国梧桐树，当时由南京市政府协助采购。西迁时只有胳膊粗细，如今已有合抱之围。江泽民曾赞叹，"这里苍松翠柏，一片青翠，环境太好了！在这样的环境里，应该出智慧，应该产生新的科学家。"

梧桐树的成长见证了交大西迁创业的艰苦磨砺与光辉荣耀，也彰显了西迁人的满腔热忱和无怨无悔。有一位老花工胡全贵，少年时便入校从事园艺工作，踊跃西迁来陕美化新校园，一辈子种花、栽树，这里的一草一木都留有他辛劳的汗水。老人家退休回老家前，在校园里走了一圈又一圈，最后抱着他亲手栽种的大树痛哭不已。正是这些普普通通的西迁人甘做"护花使者"，才成就了如今"西迁大树"的枝繁叶茂。

When autumn comes, the plane tree roads turn golden covered with fallen leaves. This is why the saying "Golden Plane Tree Roads, Poetic Jiaotong University" goes.

Over six decades having passed, the plane tree saplings have grown into towering trees, which have witnessed the hardships and glory of the westward relocation of XJTU, as well as the passion and dedication of the XJTU people. Hu Quangui, a senior gardener, was one of them, who offered to work as a gardener on the new campus when he was a young man and stayed here until he was retired. It is the hard work and efforts of tens of thousands of ordinary people like Hu that contributed to the great development of today's XJTU.

## 綠化西安新校
### 迁校工作組購置大批花木运往西安

我校一九五六年遷校工作除一些准备另星运往西安的物资外已基本結束。總务处遷校工作組目前正一方面大批購置明年用的各种傢俱和綠化西安校址的花木；一方面向各有關單位搜集資料，編造明年的搬遷計划。

目前，遷校工作組已派人前往江西、浙江和苏南各手工業城市，購置各种傢俱。同时为了綠化西安校址，已在南京、苏州、寧波、浦东等名産花木城市購置大批花木。现已运往西安的有龍柏、桂花等十多个車皮。

左上：校刊报道学校绿化　右上：西迁师生在绿化校园
中：曾经的梧桐道冬景　右下：1985年交大学子在梧桐道跑步

035

# 10.
# 西花园
位置 ｜ 腾飞塔广场西

# The West Garden
West to the Tengfei Tower Square

西园最多趣，永日自忘归。

交大人取名字都透露着一股子理工科的严谨和耿直。比如在通往图书馆的中轴线两侧的花园，便简洁明了地直呼东、西花园。这里花鸟成趣，是师生晨读静修的必选之地。

西花园是西迁精神纪念园，西安交通大学奠基者彭康校长的塑像和西迁精神纪念碑便矗立于此。

彭康是具有深厚造诣的马克思主义哲学家、久经考验的无产阶级革命家、开拓新中国高等教育的教育家。1928 年加入中国共产党。是创造社重要成员，新四军的重要领导人之一。1952 年受中央任命出任交通大学校长，次年兼任党委书记。彭康校长带领师生出色地完成了交通大学西迁重任。彭康曾首译《费尔巴哈论》（德文版）、《马克思主义与哲学》等马克思主义经典著作。

2017 年 12 月 11 日，习近平总书记对西安交大 15 位西迁老教授来信作出重要指示——"向当年响应国家号召献身大西北建设的交大老同志们致以崇高的敬意。祝大家健康长寿、晚年幸福。希望西安交通大学师生传承好西迁精神，为西部发展、国家建设奉献智慧和力量"。西迁精神纪念碑正面镌刻习近平总书记的重要指示，背面镌刻了代表西迁筚路蓝缕风雨历程的"西迁铭"。

On both sides of the central axis leading to the library are the East Garden and the West Garden with flowers and birds, where teachers and students usually read in the morning and rest.

The West Garden is a memorial garden for Westward Relocation Spirit. Here stand the statue of Peng Kang, founder and former President of XJTU, and the Monument to Westward Relocation Spirit .

Peng Kang was an educator who pioneered in the development of higher education in China , and he joined the CPC in 1928. He was an important member of the Creation Society, an important leader of the New Fourth Army of the National Revolutionary Army. He served as President of Jiaotong University from 1952 to 1968 and led the teachers and students to accomplish the westward relocation project successfully.

On December 11, 2017, Xi Jinping, general secretary of the CPC Central Committee, said in his reply to the letter from 15 senior professors who moved from Shanghai to Xi'an with the university that we should pay high respect to the faculty and staff of Jiaotong University who moved to Xi'an in response to the call of the state and devoted themselves to the construction of the northwestern China, and that he wishes all of them good health and happiness in life. He also hopes that the teachers and students of XJTU will pass on the Westward Relocation Spirit and make contributions to the development of the western region and the country as a whole. In his New Year's Address of 2018 he also pointed out that their patriotic dedication with no regrets were touching and that happiness was achieved through hard work. The front of the *Monument to Westward Relocation Spirit* in the garden is engraved with the words of Xi Jinping, and on the back is the inscription of "Westward Relocation Epigraph", describing the extraordinary journey of Jiaotong University in history.

费爾巴哈論

恩格斯著 ● 彭嘉生譯

上海南強書局版

上：彭康校长在中心楼前　　下：彭康（曾用名彭嘉生）译的《费尔巴哈论》封面　　**037**

# 11.
# 东花园

位置 | 腾飞塔广场东

## The East Garden
East to the Tengfei Tower Square

青春在此绽放，亦可重温。

东花园里有假山池塘，有草坪花树，承载了无数人的青春记忆，也留下了清晨琅琅书声、傍晚银发相携的经典画面。这里，还矗立着学校工科教育的重要奠基人唐文治和交通大学的定名人叶恭绰的塑像。

唐文治被誉为"国学大师、工科先驱"，担任校长历时 14 年，奠定学校以工为主、工管结合、工文并重的办学特色，提出"第一等"的人才培养目标，为交通大学的辉煌奠定了基础。

叶恭绰自 1920 年 8 月受命担任交通总长起即着手组建交通大学并担任校长，将交通部属的上海工业专门学校、唐山工业专门学校、北京铁路管理学校、北京邮电学校"以南洋为中坚"合并成一所学校，定名为交通大学。"五所交大是一家"，此处便是渊源。

The East Garden, featuring rockeries, ponds, lawns and flowers and trees, stand the statue of Tang Wenzhi, one of the founders of the university's engineering education, and the statue of Ye Gongchuo, who named Jiaotong University.

Tang Wenzhi, reputed as "master of Chinese traditional culture and pioneer of engineering education", served as President of Jiaotong University for 14 years and laid down the school-running characteristics of "engineering-oriented, combination of science and management, and emphasis on both engineering education and liberal arts". He also put forward the "first-class" talent cultivation goal, laying the foundation for the Jiaotong University's glory today.

Ye Gongchuo was appointed as the Director General of Transportation Department in August 1920 and set out to establish Jiaotong University and served as the President of the university. He integrated the Shanghai Special Industrial School, the Tangshan Special Industrial School, the Beiping Railway Management School, the Beiping Posts and Telecommunication School into one university with Nanyang Public College as the backbone and named it as Jiaotong University. This is the origin of the saying "Five Jiaotong Universities are of one family".

上：1986年唐文治校长塑像落成揭幕典礼　中：1991年叶恭绰塑像落成揭幕仪式　下：1996年百年校庆时的东花园喷水池　**039**

# 12.
# 腾飞塔广场
位置 | 中心楼南

## The Tengfei Tower Square
South to the Central Teaching Buildings

大鹏一日同风起，扶摇直上九万里。

1986 年，腾飞塔纪念广场落成。塔身高 21 米，塔上直冲云天、意为智慧鸟的鲲鹏高五米，寓意面向 21 世纪，五所交大比翼齐飞；塔身基座北侧以汉白玉雕刻了怀抱书本向往未来的美丽女生的塑像，被交大学子称为"不挂女神"。

广场配套改造了原已较为陈旧的图书馆门前的莲花池，改建为大型灯光喷泉。喷泉池东西两侧腰部的泵房及配电室处设立了四幅浮雕作品，刻画了交大从建校、迁校、展望等不同历史时期的奋斗场景。后为迎接百年校庆，学校在腾飞塔北侧台阶下广场东西两侧扇形并列树立了十根立柱，以铜底镌刻了交大每一个十年的发展记录。在 110 周年校庆到来之际，学校又在腾飞塔北侧广场地面绘制了中国地图，并在图上标示了五所交大的地理位置，寓意交大人在新世纪将携手奋进，为世界之光。

The construction of the Tengfei Tower Square was completed in 1986. The tower is 21 meters tall, the top of which is a 5-meter high white sculpture in the shape of a bird, known as *Kunpeng*, a Chinese legendary bird representing wisdom. The tower implies that facing the 21st century, the five Jiaotong Universities will, like the bird, be flying together side by side. To the north side of the tower on the base is a white marbled statue of a young woman with books in her arms. This young woman is known as "the goddess for success in exams", and the students are said to "worship" her whenever there is an exam.

The old lotus pond in front of the library was converted into a large colourful fountain. Four relief works were built at the pump room and power distribution room at the east and west sides of the fountain, depicting the scenes in different historical phases of Jiaotong University from its establishment, relocation to future prospect. To celebrate the 100th anniversary of Jiaotong University, ten columns were set up on the east and west sides of the square under the steps on the north side of the the Tengfei Tower, with bronze bottomed inscription on them of the development of Jiaotong University in each decade. On the occasion of the 110th anniversary of the university, a map of China was drawn on the ground of the square on the north side of the Tengfei Tower, and the geographical locations of five Jiaotong Universities were marked on the map, implying that the five Jiaotong Universities will forge ahead together in the new century to light the world.

上：1960 年代图书馆前的游泳池广场　中：改革开放初期图书馆前的荷花池

2000 年代图书馆前的腾飞塔

西安交大中心花园西区

1:250

044

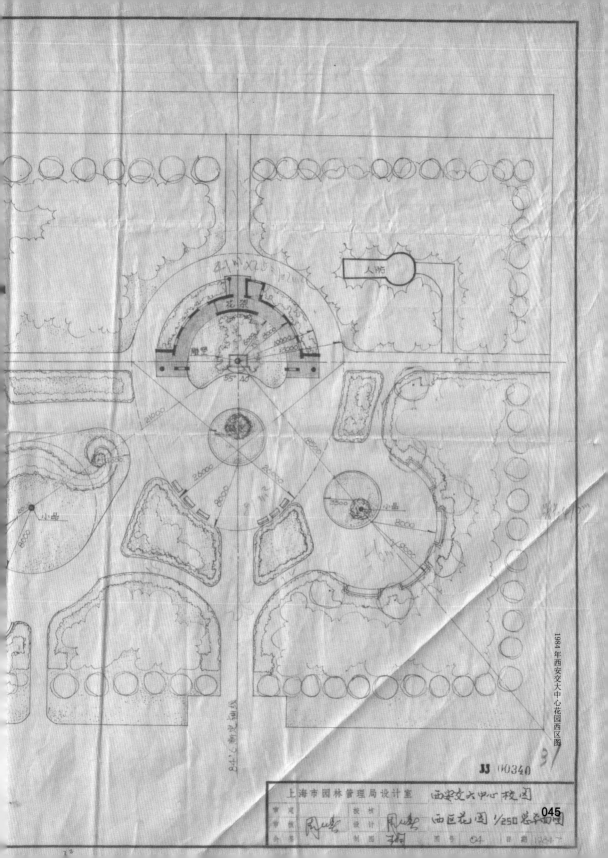

人防

花架

眺望

55°40′

小品

小曲

上海市园林管理局设计室　　西安交大中心校园

西区花园 1/250 总平面图

04　　1984.7

045

# 13.
# 钱学森图书馆
## 位置 | 四大发明广场北

# The Qian Xuesen Library
## North to the Four Inventions Square

拥书万卷日逍遥，户外绝无尘土客。

如果俯瞰图书馆，你就会发现它竟有点"机甲战士"的感觉。这里拥有藏书超过 576 万册，电子期刊 5.9 万种，电子图书 175.9 万余种，还有 24 小时"不打烊"的自习室欢迎同学们"常驻"。

从南洋公学藏书楼的血脉绵延至今，如今的图书馆于 1961 年投入使用，总建筑面积为 43794 平方米。开馆不久，便迎来了英国蒙哥马利元帅的参观考察。曾指挥过阿拉曼战役、西西里登陆、诺曼底登陆的蒙哥马利时常称赞，"这是一座亚洲一流的图书馆。"1988 年，新图书馆（南部）建成，南北二部合二为一，无缝对接。1995 年经中宣部同意，图书馆命名为"钱学森图书馆"，这也是新中国首座以人名命名的图书馆。

2009 年 6 月 5 日，国务院总理温家宝来西安交大调研，与同学们在图书馆座谈。他在图书馆门前说："希望你们立志成才，做于人民、于国家有用的人才。我们国家的前途在于提高全民族文化素质，国家的未来就寄托在年轻一代身上，你们要有决心努力奋斗！""交大是有名的学校，人才辈出。我相信在你们中间会涌现出更多的杰出人才！"临行之前，温家宝亲切地说："百年交大永远年轻，永远富有生机和活力！"

Looking in the air, the library takes a shape of somewhat like MechWarrior. The library has more than 5.76 million books, 59,000 kinds of electronic journals, and 1.759 million kinds of electronic books. Furthermore, there are also 24-hour study rooms available here.

The first library of Jiaotong University was known as Nanyang College Library. The present library was put into service in 1961 with a total construction area of 43,794 square meters. Soon after its opening, Bernard Law Montgomery, the British Marshal who directed the Battle of El Alamein, the Sicily landings, and the Normandy landings, visited the library and praised it "a first-class library in Asia". In 1988, construction of the new library (southern part) was completed and the two parts (the northern and the southern parts) were integrated into one, seamlessly connected. In 1995 the library was named "The Qian Xuesen Library" with approval of Publicity Department of the CPC Central Committee, and the then President Jiang Zemin inscribed the name of the library. It was the first library named after a person.

On June 5, 2009, Wen Jiabao, then Premier of the State Council, visited XJTU and said in front of the library, "I hope you are determined to work hard and to be equipped with abilities to serve our country and people. The future of our country lies in improving the people's qualities of the whole nation and depends on the young generation. I believe Xi'an Jiaotong University, a famous leading university, will turn out more talents. Wish this century-old university always young and full of vigor!"

一组以图书馆为封面的明信片

左上：1978 级研究生毕业留影　右上：1960 年代初同学们在图书馆前　左中：1959 年同学们在图书馆阅览室学习
右下：改革开放初图书馆夜景　左下：1990 年代同学们在图书馆前讨论

上：20世纪六七十年代同学们在图书馆前　下：1978年图书馆一角

上：1987年在建的图书馆南楼　下：1996年钱学森图书馆命名仪式

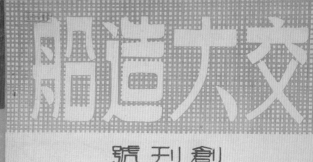
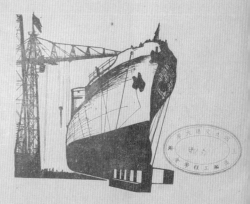

交大工程

國立交通大學經濟學會 出版

交大造船 創刊號 一九四七

交大造船工程學會

交大電機

創刊號

中華民國三十六年四月八日出版

中國電機工程師學會 編印

交大土木

第二期

中華民國三十三年十月十日出版

國立交通大學土木工程學會編印

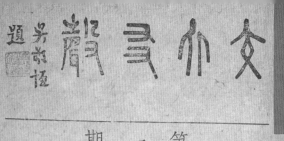

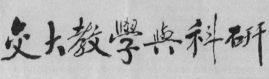

交大早期部分图书、期刊的封面

画法几何及工程制图

西安交通大学工程制图教研室 编

（上册）

科学技术出版社出版

1957

交大機械

程孝剛題

CHIAO TUNG
MECHANICAL ENGINEERING

創刊號

國立交通大學機械工程學會編印

交大月刊

交通大學學生會出版部發行

交大教學與科研

瞻閱

**7**

1958

交通大學
——多科性的重工業大學——

1896    195

交通大學
課程進度計劃

1955—1956學年第二學期

本科一年級各專業用

教務處編印
1956. 2.

交通大学
=1957=

交大早期部分图书、期刊的封面

055

# 14.
# 四大发明广场

位置 | 钱学森图书馆南

## The Four Great Inventions Square
South to Qian Xuesen Library

探索圣贤语，发明天地心。

四大发明广场又称"星农广场"，这源于学子们天马行空的想象力和联想力——四大发明，不正和英文"star farming"谐音么。

四大发明广场是在百年校庆之际落成的，包括"百年树人"纪念碑和四大发明雕塑群。

"百年树人"纪念碑下有校志一篇，记录了学校的百年变迁。一旁的西迁纪念石则书写了周恩来总理关于大西迁的批文。交通大学西迁由周恩来总理亲自领导，从决策出台到主体扎根西北，都离不开周总理的高瞻远瞩。1957 年，因沿海局势缓和，国家战略调整，交大搬迁方案形成争论，其间周总理详细了解各方面意见，并于 6 月 4 日在中南海西花厅召开专题会议讨论迁校方案，作了近万字的长篇报告。周总理指出，交大迁校及随后方案的调整应着眼于中华民族的长远发展，"一切有利于社会主义建设，一切有利于动员力量为社会主义建设服务"；方案必须坚持"支援西北方针不能变"的总原则。在周总理的亲切关怀下，交通大学西迁问题得以最终解决，从而成就了新中国高等教育史上伟大的迁徙。

四大发明雕塑群展示了印刷术、指南针、火药和造纸术。它们见证一波又一波轮滑、滑板爱好者的技压群雄，陪伴一届又一届青葱学子的步履匆匆和身着学位服的毕业留念。广场的微风，闻起来有青春的味道。

The Four Great Inventions Square was built on the occasion of the 100th anniversary of Jiaotong University, where stand Monument for A-Hundred-Year Cultivation of Talent and Sculptures of Four Great Inventions.

At the foot of the monument there is a school chronicle, recording the evolution of the university over the past century. On the Commemorative Stone of XJTU's Westward Relocation beside the monument was written a letter of approval for the Westward Relocation of Jiaotong University by former Premier Zhou Enlai, who led the whole westward relocation project.

In 1957, the relocation plan of Jiaotong University was debated due to the relaxation of coastal situation and the adjustment of national strategy. Premier Zhou, drawing on the opinions and suggestions from different aspects, made a long report at a meeting held specifically for the relocation plan in the West Flower Hall of Zhongnanhai on June 4. He pointed out that the relocation of Jiaotong University to the west and the subsequent relocation plan should be focused on the long-term development of the Chinese nation and the socialist construction to be specific, and that the relocation plan must adhere to the general principle of "the policy of supporting the development of northwest China cannot be changed". Under the leadership of Premier Zhou, the westward relocation of Jiaotong University was finally implemented, thus achieving the greatest relocation in the history of China's higher education after 1949.

The sculptures on Four Great Inventions Square show the famous ancient Chinese inventions: printing, compass, gunpowder and papermaking. The square has witnessed extraordinary skills of roller-skaters and skateboarders, and generations of students taking graduation photographs in graduation gowns.

杨秀峰部长：
八月四日高教部报告和九月四日修订报告均阅。关于交通大学迁校问题及上海、西安有关学校的调整方案，可以同意。现再正式报告批准，请即明令公布以利进行。
周恩来
一九五七.九.五

1957年周恩来总理致函高教部杨秀峰部长同意交大迁校新方案

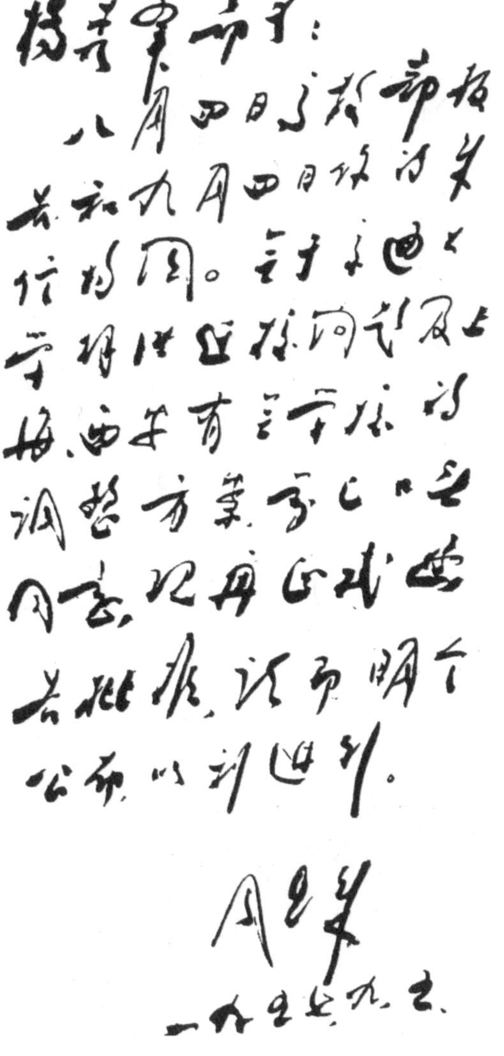

　　　上：1996年百年校庆纪念碑与四大发明雕塑落成典礼　　下：2000年中央电视台《同一首歌》走进西安交大

上：1988年逸夫科学馆奠基　　下：1989年建设中的逸夫科学馆

# 15.
# 教学主楼
位置 ｜ 文治路南

# The Main Teaching Buildings
South to Wenzhi Road

西北有高楼，上与浮云齐。愿为双鸿鹄，奋翅起高飞。

2019年10月15日，国务院总理李克强来到西安交通大学，就引用了这首《西北有高楼》。他说，有高楼就要有支撑高楼的基础，西安交大体现了"西北有高楼"。李克强总理与院士、教授们亲切交谈，对他们扎根西部培养一代代人才表示感谢，指出"你们几代人的付出让西安交大这所西北高校走在了全国高等教育的第一方阵"，希望学校为国家教育事业发展、重大科技攻关作出更大贡献。

教学主楼是兴庆校区最高楼，2006年建成投入使用，包括一体四翼共五块教学行政区。楼高23层，顶层为宣怀厅，以学校创始人盛宣怀之名命名。120周年校庆之际，盛宣怀雕像于主楼东侧塑立。

盛宣怀，清朝洋务派著名代表人物，著名政治家、企业家和慈善家，创造中华民族多项"第一"，如第一家银行——中国通商银行、第一条铁路干线——京汉铁路、第一个钢铁联合企业——汉冶萍公司等，涉及轮船、电报、铁路、钢铁、银行、纺织、教育诸多领域。其影响至大，垂及后世者，当属他秉持的"自强首在储才，储才必先兴学"理念，他先后创办了中华民族最早的两所现代大学——北洋大学堂和南洋公学，奠定了兴学育人千秋基业。

On October 15, 2019, Premier Li Keqiang visited Xi'an Jiaotong University and praised the university by quoting a poem *A Tall Building in the Northwest*. He said the "tall building" in the poem could be used to describe Xi'an Jiaotong University, building of which depended on solid foundation. Premier Li talked with academicians and professors and extended his thanks to them for their sacrifice and painstaking efforts in cultivating generations of young talents for our country. He pointed out that the hard work and efforts of generations of the faculty have made XJTU, a university in northwestern China, into the first camp of national higher education and that XJTU was expected to make greater contributions to the national educational development and major scientific and technological breakthroughs.

The Main Teaching Building, the tallest building at Xingqing Campus, was put into use in 2006. This 23-storey building is a one-body-and-four-wing architecture accommodating teaching and administration areas. The top floor is Xuanhai Hall, which was named after Sheng Xuanhuai, founder of the university. On the occasion of the 120th anniversary of Jiaotong University, the statue of Sheng Xuanhuai was erected on the east side of the Main Teaching Building.

Sheng Xuanhuai, a representative of the Westernization Movement in the Qing dynasty, a famous politician, entrepreneur and philanthropist, created many "firsts" in China, such as the first bank—Commercial Bank of China, the first railway line—Beijing-Hankou Railway, the first iron and steel joint enterprise—Hanyeping Coal and Iron Company and so on, involving many fields including shipping, telegraph, railway, steel, banking, textile and education. His most profound influence on later generations lies in the fact that he, adhering to the philosophy "A nation's self-strengthening lies in the storage of talents, which depends on development of education", successively founded China's two earliest modern universities, namely, Beiyang College and Nanyang Public College, which laid a solid foundation for the development of higher education.

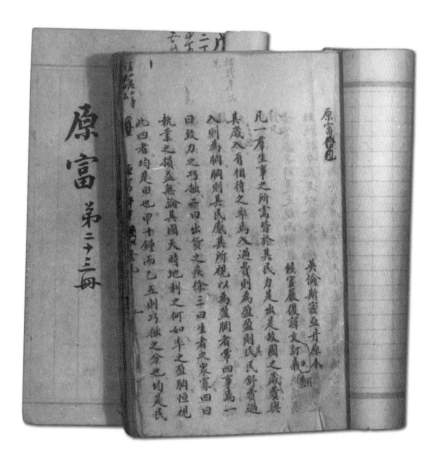

# 16.
# 孙中山铜像
位置 | 主楼南侧

## Sun Yat-sen Bronze Statue
South to the Main Teaching Buildings

大道将行，天下为公。

孙中山先生认为："交通为实业之母，铁路为交通之母"。交通大学起初致力于以铁路交通事业为基础的办学发展理念，实与孙中山关系密切。在唐文治任校长期间，孙中山曾两次来校演讲。

1912年12月，孙中山视察全国铁路返沪时来校演说，地点在上院文治堂。他首先讲了正在制定的交通建设规划，要在10年内为中国建筑铁路20万里。并殷切希望青年学子学成后投身铁路交通建设，建成一个连通全国、连通周边国家的现代化交通网络。他勉励学生：今日在校要加倍努力，发奋学习，掌握科学技术，它日才能迎头赶上，使我国与欧美发达国家并驾齐驱。

1919年五四运动后，孙中山再次应邀来校演说。他首先介绍了自己新近完成的《实业计划》一书，提出了铁路建设思想和规划全国铁路建设的庞大计划，向师生们描绘了一幅从交通建设方面实现中国现代化的宏伟蓝图。他还介绍了三峡水力发电、改善长江航运的设想。其后，孙中山为本校杂志《南洋》题词："强国强种"。凌鸿勋在场聆听演说，感受颇深，立志从事交通工程技术，后成为我国土木工程专家。

这座孙中山先生铜像是2011年5月27日矗立在交大校园的，时逢纪念辛亥革命100周年。铜像位于校园中轴线上，底座正面镌刻"天下为公"四个大字，背面则是孙中山先生手书《礼运·大同篇》全文。

Dr. Sun Yat-sen believed that transportation breeds industry and railway breeds transportation. So Jiaotong University at the beginning stage devoted to be running based on railway transportation, which was closely related to Sun Yat-sen. During Tang Wenzhi's tenure as president, Sun Yat-sen came to the university twice and delivered speeches.

In December 1912, Dr. Sun Yat-sen gave a speech at Wenzhi Hall at Shangyuan when he returned to Shanghai after his inspection of railway construction around the country. He talked about the railway construction plan being worked out, which was to build about 62,200 miles railways in China within 10 years. He hoped that young students could join in railway transportation construction after graduation to build a modern transportation network covering the whole country and connecting our neighboring countries. He also encouraged students to study hard and learn scientific and technological skills in order for China to catch up with the developed countries in Europe and America.

After the May 4th Movement in 1919, Dr. Sun Yat-sen was invited to give a speech at the university again. He introduced his newly completed book *The International Development of China*, in which he put forward his ideas of railway construction and the grand plan of constructing national railway network, and he described the modernization blueprint of China through transportation construction from railway construction to building ports. He also introduced his idea of water conservancy and power generation at the Three Gorges to improve navigation on the Yangtze River. Finally, he inscribed "Qiangguo Qiangzhong" (a strong country with strong seedlings) for *Nanyang*, a periodical of the university. Deeply touched and excited after listening to his speech, Ling Hongxun, who graduated from the Shanghai Special Industrial School, was determined to engage in the study of transportation engineering technology and later became an expert in civil engineering in China.

The bronze statue of Sun Yat-sen was erected on the central axis of the campus on May 27, 2011, the 100th anniversary of the 1911 Revolution. On the front of the base, four Chinese characters of "Tian Xia Wei Gong" (The world is for all.) are carved. On the backside of the base is Dr. Sun's handwriting of the full text of an ancient essay "The World of Dah-Torng" (a Great Utopia)

強國強種

孫文題

# 17.
# 思源学生活动中心
位置 | 南门内

# Siyuan Student Activities Center
Close to the South Gate

问渠那得清如许？为有源头活水来。

"海纳百川"的气量，赋予了这里多重的气质。时而庄严，校歌、权杖、拨穗、颁证，一届届学子身着学位服聆听师长最后一课，极富仪式感的毕业时刻承载着太多的祝福和期待；时而青涩，一抹抹迷彩绿在这里汇报军训成果，一张张青春面庞在开学典礼上憧憬大学时光；时而激烈，"必胜"的口号响彻球场，胶着的比分紧紧吸引着场下的目光，这是属于篮球的高光时刻；时而典雅，急管繁弦、声振林木，让人流连在音乐中"沉醉不知归路"；时而酷炫，各类赛事、相声、演唱会、发布会、讲座等丰富着校园里的生活。

场地面积近万平方米的思源学生活动中心建设于百年校庆之际，馆名为江泽民学长来校视察之际所题写。

Siyuan Student Activities Center, covering an area of nearly 10,000 square meters, was built on the occasion of the 100th anniversary of the university, and its name was inscribed by Jiang Zemin, former President and the university's graduate when he visited the university.

Siyuan Student Activities Center is where various activities are held and staged, ranging from graduation commencement, demonstration of military training results, basketball games, concerts, lectures, to different types of contests, etc.

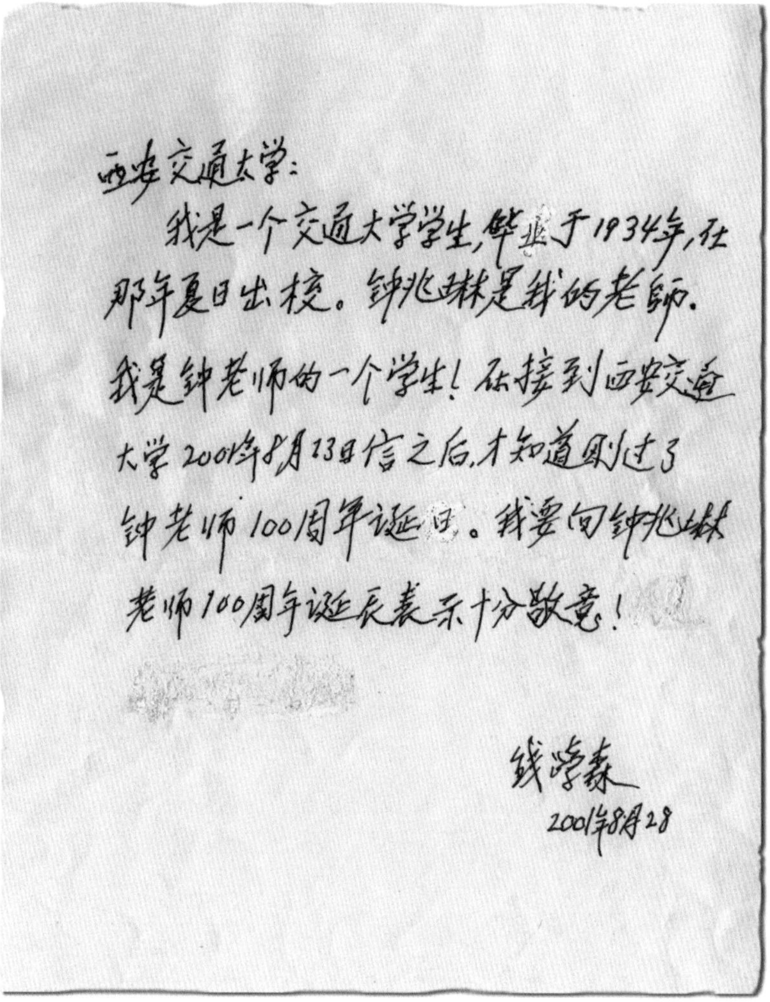

西安交通大学：

我是一个交通大学学生，毕业于1934年，在那年夏日出校。钟兆琳是我的老師。我是钟老师的一个学生！在接到西安交通大学2001年8月13日信之后，才知道刚过了钟老师100周年诞日。我要向钟兆琳老师100周年诞辰表示十分敬意！

钱学森

2001年8月28

2001年钱学森学长贺老师钟兆琳先生诞辰100周年信件

上：1996 年思源学生活动中心命名典礼　下：1998 年中国大学生篮球联赛在思源学生活动中心举行

上：1999 年 5 月 3 日，为庆祝五四运动八十周年，1999 名学生与嘉宾共同演唱《黄河大合唱》
下：2000 年 4 月 17 日，西安交通大学、西安医科大学、陕西财经学院合并大会

# 18.
# 交大西迁博物馆
位置 ｜ 思源学生活动中心东

# Jiaotong University Westward Relocation Museum
East to Siyuan Student Activities Center

忆昔西征日，飞腾尚少年。

要了解交大西迁和西迁精神，必去打卡地就是交大西迁博物馆了。交大西迁博物馆于 2018 年 12 月 11 日面向社会正式开放，占地面积约为 940 平方米，建筑面积约为 3760 平方米，共 4 层，采用全钢结构建设。馆内布展面积 2400 平方米，展线全长 699 米，由序厅、放映厅、展厅和多功能厅组成，内容分为溯源、西迁和致远三个部分，展出照片、图表、实物等 2200 余件，集中体现西迁人波澜壮阔的创业历程和辉煌成就，展示西迁精神激励一代代知识分子奋勇前进的磅礴伟力。西迁博物馆获批陕西省爱国主义教育基地、陕西省社会科学普及基地等，被评为陕西省"学雷锋活动示范点"。

2020 年 4 月 22 日，习近平总书记在西安交通大学考察调研时指出，西迁精神的核心是爱国主义，精髓是听党指挥跟党走，与党和国家、与民族和人民同呼吸、共命运，具有深刻现实意义和历史意义。习近平勉励广大师生大力弘扬西迁精神，抓住新时代新机遇，到祖国最需要的地方建功立业，在新征程上创造属于我们这代人的历史功绩。

To learn about the Westward Relocation and its spirit, you must visit the Westward Relocation Museum of XJTU. It was officially opened to the public on December 11, 2018. The Museum, constructed with all-steel structure, covers an area of about 940 square meters, with a building area of about 3,760 square meters and a total of 4 floors. The exhibition area of the museum is 2,400 square meters, with the exhibition length totaling 699 meters. It consists of preface hall, projection hall, exhibition hall and multi-function hall, presenting three major parts of contents: tracing back to its origin, westward relocation and looking into the future. More than 2,200 photos, charts and objects are displayed, showing the extraordinary pioneering history and brilliant achievements of those who moved westward to Xi'an, as well as the power of westward relocation spirit inspiring generations of XJTU people to forge ahead bravely. Westward Relocation Museum was granted Shaanxi provincial-level patriotic education base and Shaanxi provincial-level social science popularization base, and is approved to be "Demonstration Area of Learning from Lei Feng".

Xi Jinping, general secretary of the CPC Central Committee, visited XJTU on April 22, 2020 and pointed out that the Westward Relocation Spirit was based on a sense of patriotism with the essence of sharing the destiny with the nation and the people, which is of profound practical and historical significance. He also encouraged teachers and students of XJTU to carry forward the Westward Relocation Spirit, seize the opportunities in the new era, pursue their careers in the most-needed places of the country to make new contributions and create new achievements.

上：西迁时期的算盘　下：钟兆琳先生使用过的物品

　　　上：西迁时学校办公电话　　下：见证西迁历程的英文打字机

上：出诊包　下：交大摩托车训练班使用过的摩托车

　　　　上：西迁时期基础实验器材　　下：西迁时期电子管收音机、棉大衣

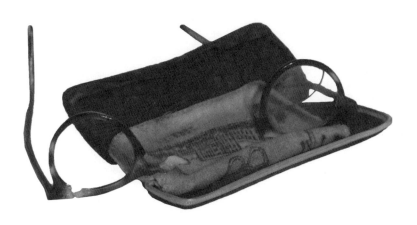

上：西迁时期教学用照相机　下：钟兆琳先生使用过的物品

# 19.
# 博物馆
位置 | 校园西南角

## The Museums
Southeast Corner of the Campus

谁知石烂山枯后，犹有残碑记汉唐。

来交大博物馆，带你秒回周秦汉唐。这里拥有4500余平方米的陈列面积，历代艺术文物馆、西部农民画馆、木版年画馆、书法碑石馆、陕西秦腔博物馆、邢良坤陶瓷艺术馆等多个陈列馆罗列其中。馆藏文物及珍贵艺术品4900余件，见证时代的变迁。著名书法家、西安交通大学博物馆名誉馆长钟明善先生说，"文化艺术是人民创造的，所以就应该用之于民，服务社会。"博物馆内近3000件藏品都源于他的个人捐赠。

你知道穿越秦砖汉瓦的百戏之祖是秦腔吗？你知道秦人歌唱《诗经》中的"秦风"正是秦腔的起源吗？快来秦腔馆感受一场视听的饕餮盛宴吧，这里的秦腔项目早已是教育部第一批中华优秀传统文化传承基地了。《王翊元夫妇合葬墓志》是如今唯一可以看到的李商隐书法作品，还有韩择木所书《大唐寿光公主墓志》以及徐浩所书"神道碑"，热爱人文艺术的你一定不能错过碑石书法展馆。西汉壁画墓VR展示可以带你一睹中国最早的二十八宿古天象图，有如置身于在渭河两岸挺立起来的西汉王朝……

走吧，让我们再往时光深处进发，重回盛世长安！

A visit to XJTU Museums will bring you back to thousands of years ago right away.

The museums have an exhibition area of more than 4,500 square meters, including Chinese Cultural Relics and Art Gallery, the Western China Peasant Painting Gallery, the Woodblock New Year Painting Gallery, the Stele Calligraphy Gallery, Museum of Qinqiang Opera in Shaanxi Province, and the Xing Liangkun Ceramic Art Gallery. There are more than 4,900 cultural relics and precious works of art in the museums, witnessing the changes of time. Mr. Zhong Mingshan, a famous calligrapher and honorary curator of XJTU Museums, said, "Culture and art are created by the people, so they should be used for the people and for the society." Nearly 3,000 collections in the museums are from his personal donation.

If you want to learn about the origin of Qinqiang Opera, known as the originator of Chinese operas, the Qinqiang Opera Museum cannot be missed, which has been granted as among the first batch of China's excellent traditional culture inheritance bases by the Ministry of Education. If you love humanities and arts, come and visit the Stele Calligraphy Gallery, which houses *Epitaph of Wang Yiyuan and His Wife*, the only calligraphy work of Li Shangyin's, a famous poet in the Tang dynasty, and *Epitaphs of Princess Shouguang* of the Tang dynasty by Han Zemu, calligrapher of the Tang dynasty, and the Chinese characters "Shen Dao Bei" (tablet on the side of a tomb giving biographical sketch) inscribed by Xu Hao, calligrapher of the Tang dynasty. In addition, in the VR display of the Western Han Dynasty Mural Tombs can give you a glimpse of China's earliest "Twenty-eight Constellations" of night sky picture.

Come on! Let's travel back in time to the prosperous Chang'an of the Tang dynasty.

韩择木书《大唐故寿光公主墓志铭》（局部）

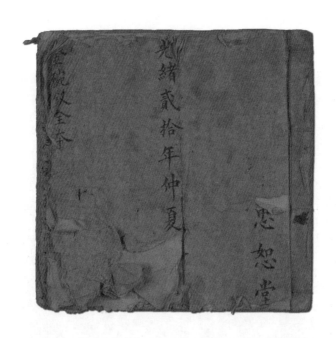

皮影剧本《金碗钗》

皮影 穆桂英

# 穿越路线二
## 历史文脉散步

# 穿越须知

## What You Must Know
## Before Starting the Journey

　　欢迎大家回来！刚刚我们进行了六十多年前的穿越之旅，体验了那个激情燃烧的西迁岁月。现在我们要穿越千年，去感受汉唐长安的盛世气象。

　　当我们穿越到西汉时期，你会发现我们正处在汉长安城东南方的上林苑中。汉武帝时的一代大儒董仲舒和汉宣帝时的名臣萧望之都长眠于此。当我们穿越到唐代长安，你会发现我们脚下的交大校园正是唐长安城道政、常乐两坊之所在。你会听到校园北门外兴庆宫中唐玄宗和杨贵妃听奏《霓裳羽衣曲》的飘渺乐声；漫步道政、常乐两坊，你会看到坊街边宏伟幽深的和政公主府，还有开国功臣侯君集的府邸，甚至也会发现一代枭雄安禄山的宅第。在这里，你可能会遇到一代名将郭子仪来探访父亲郭敬之的车马仪仗，也可能会看见大诗人王维从哥哥王缙的府第中踏月而归，还可能和刚刚在东亭寓所中策马而出的诗人白居易不期而遇。夜色将深，你可能会听到动人心弦的琵琶女的弹奏，还有名闻一时的尉迟将军的美妙的筚篥曲声。在道政坊门畔，你还可能会目睹唐代传奇小说《柳氏传》中韩翃等待柳氏赴约时的焦急与不安；你也可以逢人询问，在这里出土的三重银盒和玉花簪头的主人是谁……

　　开启新的穿越之旅吧！

Welcome back! We have just taken a journey back to more than 60 years ago, experiencing the burning years of moving west. Now we have to travel through thousands of years to feel the atmosphere of Chang 'an in Han and Tang dynasties.

When we travel to the Western Han dynasty, you will find that we are in the Shanglin Garden in the southeast of Chang 'an City. Dong Zhongshu, a great scholar of Emperor Wudi of the Han tynasty, and Xiao Wang, a famous official of Emperor Xuandi of the Han tynasty, were buried here. When we travel to Chang 'an of the Tang tynasty, you will find that the campus of Jiaotong University at our feet is exactly where Dao Zheng of Tang tynasty and Changle Square are located. You will hear the ethereal music of Tang Xuanzong and Yang Guifei in the Xingqing Palace outside the north gate of the campus playing *"Song of Neon Clothes and Feather Clothes"*. Take a stroll through the two alleys of Dao Zheng and Chang Le, and you will see the grand and deep palace of He Zheng Princess, the mansion of the founding hero Hou Junji, and even the mansion of an Lushan. Here, you may encounter general Guo Ziyi to visit his father Guo Jingzhi's horse in a guard of honour, may also see Wang Wei returned from his elder brother Wang Jin's mansion at night, still may unexpectedly encounter in the east pavilion apartment the poet Bai Juyi. It will be late at night and you may hear the pipa lady playing and the nice Tartar pipe of General Yuchi who was famous for a while. At the gate side of Dao Zheng Fang, you may also see the anxious Maiden Liu go for date with poet Han Hong described in a legendary novel *Stories of Maiden Liu and Han Hong*, being anxious and nervous; You can also meet people to ask who is the owner of the three silver boxes and jade flower hairpins unearthed here...

Let's start a new time travel!

# 一.
# 兴庆宫
位置 | 北门北

## The Xingqing Palace
North to the North Gate

香车宝马嬉游盛，别馆离宫往事非。

西安交通大学北边的邻居，就是开创开元盛世的唐玄宗李隆基。大足元年（701年），临淄郡王李隆基与四位兄弟宁王李宪、申王李㧑、岐王李范及薛王李业，一同被武则天赐宅于隆庆坊，号称"五王宅"。后因避玄宗名讳，改为兴庆坊。开元二年（714年），李宪奏请献兴庆坊宅为离宫，即兴庆宫。开元十六年（728年），唐玄宗从大明宫移仗兴庆宫听政。上元元年（760年），已经退位的唐玄宗被逼离开兴庆宫。

当时的兴庆宫东西宽1080米，南北长1250米，总面积为1.35平方千米，比现存的明清故宫还要大出一倍。

在兴庆宫遗址上建设的兴庆公园占地780亩。公园内的兴庆湖是在唐代"龙池"原址上挖掘的人工湖。据史载，龙池是皇帝宴乐游赏之地。公元710年，这里曾举行轻舟竞赛大会，当时诗人李适曾有"轻舸白帆迅疾，冲开南山倒影"的咏叹。20世纪50年代初，西安市规划局在交大西迁选址时已有兴庆公园复建规划，其后交大师生利用课余时间参与了兴庆公园建设，许多西迁人都对挖掘兴庆湖记忆犹新。

North to Xi'an Jiaotong University was the residence of Li Longji, Emperor Xuanzong of Tang dynasty, who started the golden age of Kaiyuan. In 701, Li Longji, Prince of Linzi, and his four brothers, Li Xian (Prince of Ning), Li Wei (Prince of Shen), Li Fan (Prince of Qi), and Li Ye (Prince of Xue), were enfeoffed residences by Wu Zetian (the empress) in Longqing Fang (Fang refers to the living quarters), known as "Mansion of Five Princes". The name of Longqing Fang was later changed into Xingqing Fang to avoid using the same character "Long" as in the Emperor's given name. In 714, Li Xian petitioned to present the mansion in Xingqing Fang to the Emperor Xuanzong, which was transformed into Li Palace, known as the Xingqing Palace. In 728, Emperor Xuanzong moved from Daming Palace to Xingqing Palace to administer state affairs. After An-Shi Rebellion (the rebellion was headed by An Lushan and Shi Siming, treacherous court officials; hence the name), the Xingqing Palace lost its importance in politics and became a palace where Emperor Xuanzong lived after his abdication in 760.

With 1,080 meters in width from east to west and 1,250 meters in length from north to south, the Xingqing Palace then covered a total area of 1.35 square kilometers, twice as large as the existing Imperial Palace of Ming and Qing dynasties.

The Xingqing Park built upon the site of the Xingqing Palace covers an area of nearly 52.3 hectares, and is the largest urban park in Xi'an. The man-made Xingqing Lake in the the Park was built on the site of the original "Dragon Pool" of the Tang dynasty. According to the historical records, the "Dragon Pool" was specially built for entertainment, where Emperor Xuanzong used to go boating, sightseeing and feasting together with his concubines and officials. In 710, a boating race was held here, which was depicted in a poem by Li Shi, a poet of the Tang dynasty. In the early 1950s when Jiaotong University was about to move to Xi'an, Xi'an City Planning Bureau planned to build the Xingqing Park and the faculties and students of Jiaotong University all joined in the construction of the Park in their spare time during that period, the memory of which remains fresh to many of those involved.

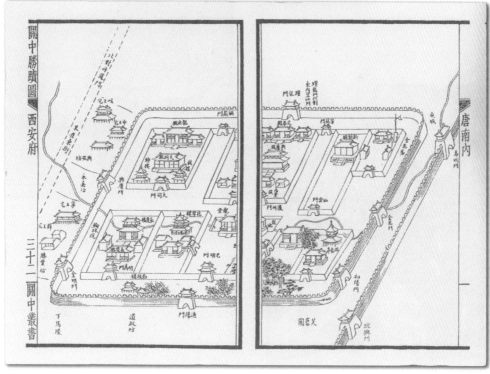

上：兴庆池公园规划示意图　下：唐南内（兴庆宫）图

西安兴庆公园留影 '69

一组兴庆公园杂忆

兴庆宫公园 1981

兴庆公园杂忆

西安市兴庆公园
壹角游泳人场券
壹角 每票一人 隔场作废 场次

西安市兴庆宫公园水上活动收费票
持票 0402490 过时
乘坐 陆 角 不坐作废
船车
上下船
时间 由 时 分至 时 分

西安市兴庆宫公园
副卷 小火车票 只限一人
贰角

西安兴庆宫公园自控飞碟
每票只限一人 游艺活动票 每场五分钟
票价 伍角

碰碰车
兴庆宫公园儿童游乐场
票价: 1.20元

兴庆公园钓鱼票
依作审定儿
票价
编号 0049875
时间 由上午 时 分至 时 分 钓鱼区
注事 1.不得越过钓鱼区
壹事 2.听从工作人员指导
券

激光枪游艺票
存根(一张)
No.0310299

西安市兴庆宫公园
激光枪游艺票
每次十枪 票价: 捌角 每票一人

准说是难题，完全是必题。
我们曾相识，我能也共过。
兴英湖畔的歌声，隆冬飞梭的雪球
更记忆。

华山道上的展形会。
结下了姐弟情谊。

你。
待人接物，很有气派。
谦逊好学，智慧超群。

但祝愿。
今日高飞，体育颂传。
侨乡更阔，听君前展。

兴庆公园杂忆

089

一组兴庆公园杂忆

一组兴庆公园杂忆

　　　　　　　　兴庆公园杂忆

兴庆公园杂忆

# 二.
# 唐侯君集宅
位置 | 中心楼北

## Site of Residence
## of Hou Junji of the Tang Dynasty
North to Central Teaching Buildings

请君试上凌烟阁，若个书生万户侯。

侯君集（？—643年），豳州三水县（今陕西旬邑）人，唐朝名将，官至宰相、兵部尚书。侯君集少年时即以勇武著称，曾随李世民东征西战，功勋卓著。在玄武门之变中，侯君集发挥了重要作用。唐武德九年（626年），李世民与太子李建成的矛盾激化。侯君集与尉迟恭劝谏李世民早下决心，支持李世民发动玄武门事变。

在玄武门之变中，侯君集负责率军控制唐高祖和朝臣，尉迟恭等人则诛杀李建成及李元吉全家。贞观八年（634年），侯君集作为副将随李靖一起讨伐吐谷浑，进献奇计，大破敌军。贞观十四年（640年），唐太宗命侯君集为行军大总管，率军征讨西域高昌国，大获全胜。不过，侯君集后因附逆太子李承乾意图谋反，被太宗处死。

侯君集也是凌烟阁二十四功臣之一。凌烟阁位于太极宫三清殿旁。唐朝贞观十七年（643年），唐太宗为纪念一同打天下的诸多功臣，命画师阎立本在凌烟阁内描绘了二十四位功臣的画像。阁中画像分为三层：最内一层为功勋最高的宰辅之臣，中间一层为功高王侯之臣，最外一层则为其他功臣。画像的比例皆依真人大小，均面北而立，唐太宗时常前往凌烟阁缅怀旧日功臣。这二十四位功臣是：赵国公长孙无忌、河间王李孝恭、莱国公杜如晦、郑国公魏徵、梁国公房玄龄、申国公高士廉、鄂国公尉迟敬德、卫国公李靖、宋国公萧瑀、褒忠公段志玄、夔国公刘弘基、蒋国公屈突通、郧节公殷开山、谯国公柴绍、邳国公长孙顺德、郧国公张亮、潞国公侯君集、郯国公张公谨、卢国公程知节、永兴公虞世南、刑国公刘政会、莒国公唐俭、英国公李勣和胡国公秦琼。

Hou Junji was a well known general of the Tang dynasty and served as the Minister of War. Famous for his bravery and command of strategies, Hou Junji played a critical role in Xuanwu Gate Incident (a palace coup), which ensured that Li Shimin, Prince of Qin, became Empror Taizong of Tang. He also made great achievements in destroying enemies including Tuguhun and Gaochang by contributing surprising plans. But it's a pity that Hou was eventually implicated in a plot by the crown prince, Li Chengqian, to overthrow Emperor Taizong, and was executed.

Hou Junji was also one of the 24 meritorious officials at Lingyan Pavilion beside Sanqing Hall in Taiji Palace. In 643, Emperor Taizong commissioned the artist Yan Liben to paint life-sized portraits of 24 officials to commemorate them for their meritorious services and extraordinary contributions to the founding of Tang dynasty.

附
大士三樊

凌煙閣　劉源敬繪

吳門桂笏堂

吏部尚書陳國公侯君集

功名圖麒麟戰

肯當遠杉
董文敏
濠

凌烟阁功臣图之侯君集（清刘源敬绘）

# 三.
# 唐代三重银盒出土地
位置 | 西二楼

# Excavation Site
# of Triplex Silvery Box of the Tang Dynasty
West Building 2

钗留一股合一扇，钗擘黄金合分钿。

1979 年 9 月 24 日，西安交大校园西北侧基建时，出土了三件套装在一起的唐代银盒。根据纹饰的不同，银盒分别称为：都管七国六瓣银盒、鹦鹉纹海棠形圈足银盒、龟背纹银盒。龟背纹银盒内还装有水晶珠两颗、玛瑙珠一颗。

都管七国六瓣银盒正中的六角形内，錾刻一个骑象人，象身备有鞍鞯。前有顶物膜拜者，后有持伞盖者，表明骑象人的身份很尊贵。象的右侧有一人站立，左侧有一人随行，一人随地而坐。象的左上方有"昆仑王国"题榜。从中央的昆仑王国右侧起，顺时针排列有以下诸国：婆罗门国、土番国、疏勒国、高丽国、白柘羯国和鸟蛮人。

银盒中所提到的七个国家，除"昆仑王国"和"白柘羯国"外，皆在《唐书》中有传。这套三重银盒明显带有中亚、西亚的艺术风格，是唐代丝绸之路繁荣的见证。该银盒为国家一级文物，现藏于西安博物院。

On September 24, 1979, a triplex silvery box of the Tang dynasty was unearthed on the northwest side of the campus of Xi'an Jiaotong University, each piece with a different pattern. The biggest one (the outer one) is about Governance of Seven Nations. The one in the middle is decorated with parrots with crabapple shaped round foot; the smallest one (the inner one) has a pattern of back of a turtle's shell with two crystal beads and one agate bead inside it.

The central part of the lid of the outer silver box is a hexagon, in which a man on an elephant with a saddle is carved. There are worshipers in front holding gifts on their heads and umbrella holders at the back, indicating the elephant rider is a noble man. On the right side of the elephant stands a man, and on the left side, a man follows and the other sits on the ground. On the upper left of the elephant is an inscribed board which writes "Kunlun Kingdom". Starting from the right side of the "Kunlun Kingdom" in the center, the following states are arranged clockwise: Brahman state, Tuban State, Shule State, Koryo State, White Zhejie Kingdom, and Nanzhao State. The seven states mentioned above were all recorded in *The Book of Tang* except "Kunlun Kingdom" and "White Zhejie Kingdom." This triplex silver box has distinct artistic style of central and western Asia, which bears witness to the prosperity of the Silk Road in the Tang dynasty. It is recognized as a national first-class cultural relic and is now collected in Xi'an Museum.

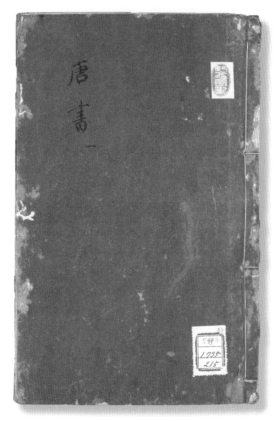

上：西安交大出土都管七国六瓣银盒（现藏西安博物院）　下：《新唐书》书影（顺治十三年汲古阁本）

# 四.
# 唐代梵文石碑出土地
位置 | 四大发明广场

/

# Excavation Site
# of Sanskrit Stele of the Tang Dynasty
Four Great Inventions Square

鹤立蛇形势未休，五天文字鬼神愁。

1985 年，西安交大在图书馆建筑工地发掘出梵文咒语石碑一块，碑高 12.5 厘米，宽 42.5 厘米，厚 8.4 厘米。碑上横书古梵文七行，为驱鬼降妖、保佑平安之内容。碑面中央镌以佛家法器图形，四侧饰以卷云纹图案。现藏于西安交通大学博物馆。

梵语是印度的古典语言，是印欧语系最古老的语言之一，也是印度、中亚很多种语言的祖先。因为古人相信其为梵天所造，故称"梵语"。梵语由 48 个字母组成，包括 34 个辅音和 14 个元音。梵语是印度国家法定的 22 种官方语言之一，但却是使用人数最少的语言，和拉丁文一样，不再是日常生活的交流语言。耆那教、佛教、印度教的很多典籍和文学作品都是通过梵文保存下来的。因此，汉传佛教的很多经典都是从梵文翻译而来的。古代高僧法显、玄奘入印度求法时，看到的都是梵文所书写的佛教经典。

In 1985, a stele inscribed with incantation in Sanskrit (ancient Indian) was excavated at the library construction site in Xi'an Jiaotong University. The stele is 12.5cm high, 42.5cm wide, and 8.4 cm thick. Seven lines of ancient Sanskrit are transversely inscribed on the surface of the stele to drive away ghosts and pray for safety. The central part of the stele is engraved with the patterns of Buddhist Ritual Tools, with cirrus cloud patterns around it at the four sides. The stele is now collected in the Museums of Xi'an Jiaotong University.

Sanskrit is the classical Indian language and one of the oldest languages in Indo-European language family, yet it is no longer the language for daily communication like Latin. Many Buddhist and Hindu classics and literary works are preserved in Sanskrit and many Chinese Buddhist classics are translated from Sanskrit.

唐代梵文石碑

# 五.
# 唐张平高宅、张行成宅
位置 ｜ 钱学森图书馆南

## Zhang Pinggao's Residence
## and Zhang Xingcheng's Residence of the Tang Dynasty
South to Qian Xuesen Library

圣代贤才萃一时，公侯将相总相宜。

张平高（生卒年不详），绥州人（今陕西绥德）。隋朝末年任职鹰扬府校尉，戍守太原，与唐高祖李渊结交，参与起义。李渊登基后，张平高官拜左领军将军，被封萧国公。武德初年（618年），唐高祖下诏奖赏太原起兵的元勋，张平高受到特赏，即一生中可以免除一次死罪。唐武德九年（626年）十月，唐太宗即位，再次论功行赏，封张平高食邑300户。太宗贞观初年（627年），张平高出任丹州刺史，因故病免，以右光禄大夫之职返乡，死后被封为罗国公，追赠潭州都督。史载张平高宅于道政坊南门之西第二家，即现钱学森图书馆前。

张行成（587—653年），字德立，定州义丰（今河北安国）人，唐高宗时宰相。张行成隋末入仕，唐初登进士科，授富平主簿，又转殿中侍御史。中正不阿，经常直谏太宗。贞观十九年（645年），太宗亲征高丽，张行成等五人共同辅佐皇太子。太宗驾崩后，张行成与高季辅共同辅佐高宗即位，以顾命大臣辅政。高宗永徽二年（651年），拜尚书左仆射。永徽三年（652年），高宗立陈王李忠为太子，张行成兼太子太傅。永徽四年（653年），张行成卒于官舍。

Zhang Pinggao, born in Suizhou (today's Suide, Shaanxi province), became friends with Li Yuan in the uprising at the end of the Sui dynasty. When Li Yuan became Emperor Gaozu and ascended the throne, Zhang Pinggao was appointed as a general and was honored as Duke of Xiao. In October, 626, Emperor Taizong of Tang ascended the throne and greatly rewarded Zhang Pinggao based on his performance and achievements. According to the historical records, his residence was the second house in the west of the Southern Gate of Daozheng Fang, south to today's Qian Xuesen Library.

Zhang Xingcheng (587 – 653) served at different important positions in the Tang dynasty and played an important role during the reigns of Emperor Taizong and his son Emperor Gaozong of Tang. He was upright and straight, and often admonished Emperor Taizong directly. Zhang became the Prime Minister of Tang during the reign of Emperor Gaozong. He was also appointed as the Grand Tutor of the Crown Prince (*Taifu*), teaching him reading, writing and learning other skills.

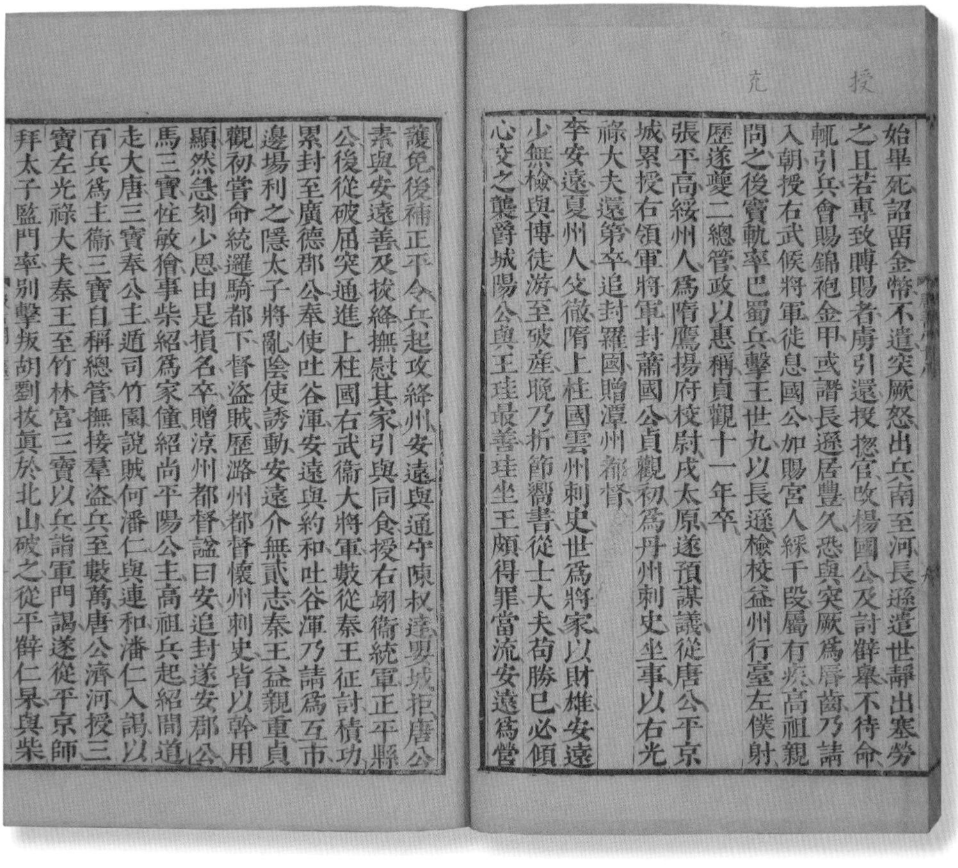

始畢死詔賻金幣不遣突厥怒出兵南至河長遜遁世靜出塞勞
之且若專致賻賜者虜引還授揔官改楊國公及討辭舉不待命
輙引兵會賜錦袍金甲或許長遜居豐久恐與突厥爲屑藚乃請
入朝授右武候將軍徙息國公加賜宮人緜千段屬有疾高祖親
問之後寶軌率巴蜀兵擊王世九以長遜檢校益州行臺左僕射
歷遂夔二總管政以惠稱貞觀十一年卒

張平高綏州人爲隋鷹揚府校尉戍太原遂預謀議從唐公平京
城累授右領軍將軍封蕭國公貞觀初爲丹州刺史坐事以右光
祿大夫還第卒追封羅國贈潭州都督

李安遠夏州人父微隋上柱國雲州刺史世爲將家以財雄安遠
少無檢與博徒游至破産睍乃折節嚮書從士大夫苟勝已必傾
心交之襲爵城陽公與王連最善珪坐王頍得罪當流安遠爲管

護兒後補正平令以兵起攻絳州安遠與通守陳杰遠出城拒唐公
素與安遠善及拔絳撫慰其家引與同食授右胡衛統軍正平縣
公後從破屈突通進上柱國右武衛大將軍敕從秦王征討積功
累封至廣德郡公奉使吐谷渾安遠與約和吐谷渾乃請爲互市
邊場利之隱太子將亂陰使誘動安遠介無武志泰王益親重貞
觀初嘗命統羅騎都下督盜賊歷潞州都督懷州刺史皆以幹用
顯然急刻少恩由是損名卒贈涼州都督追封安郡公

馬三寶性敏猾事柴紹爲家僮說紹盜兵起尚平陽公主迎兵起
走大唐三寶奉公主遣司竹園說賊何潘仁與連和潘仁入謁以
百兵爲土籍三寶自稱總管撫接羣盜兵所載萬唐公濟河授三
寶左光祿大夫秦王至竹林宮三寶以兵詣軍門謁遂從平京師
拜太子監門率別擊阪胡劉拔眞於北山破之從平辭仁杲與柴

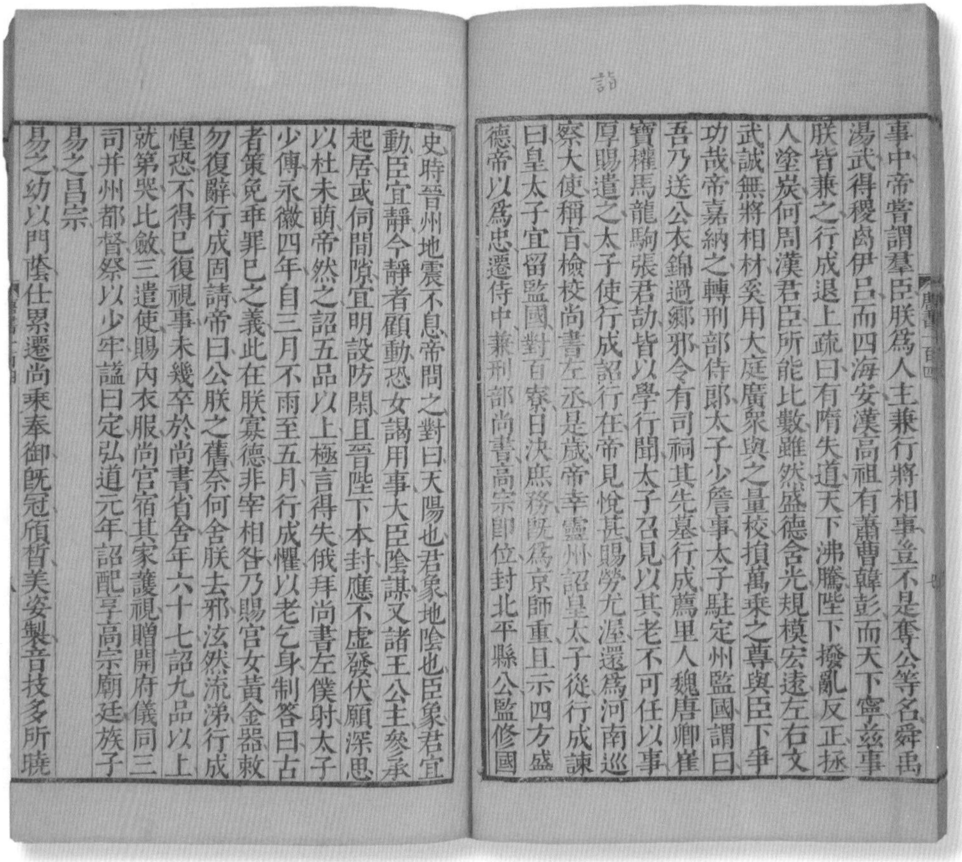

詔

事中帝嘗謂羣臣朕爲人主兼行將相事豈不是奪公等名舜禹
湯武得稷卨伊呂而四海安漢高祖有蕭曹韓彭而天下寧茲事
朕皆兼之行成退上疏曰有隋失道天下沸騰陛下撥亂反正拯
人塗炭何周漢君臣所能比數雖然盛德含光規模宏遠左右文
武誠無將相材奚用大庭廣衆與之量校損萬乘之尊與臣下爭
功哉帝嘉納之轉刑部侍郎太子少詹事太子駐定州監國謂曰
吾乃送公衣錦過鄉邪令有司祠其先墓行成薦里人魏唐卿崔
寶權馬龍駒張君劼皆以學行聞太子召見以其老不可任以事
厚賜遣之太子使行成詔行在帝見悅甚賜勞尤渥還爲河南巡
察大使稱旨檢校尚書左丞是歲帝幸靈州詔皇太子從行成諫
曰皇太子宜留監國對百寮日決庶務既爲京師重且示四方盛
德帝以爲忠遷侍中兼刑部尚書高宗即位封北平縣公監修國

唐書一百四

史時晉州地震不息帝問之對曰天陽也君象地陰也臣象君宜
動臣宜靜今靜者顧動恐女謁用事大臣陰謀又諸王公主參承
起居或伺間隙宜明設防閑且晉陛下本封應不虛發伏願深思
以杜未萌帝然之詔五品以上極言得失俄拜尚書左僕射太子
少傅永徽四年自三月不雨至五月行成懼以老乞身制荅曰古
者策免乖罪己之義此在朕寡德非宰相咎乃賜宮女黃金器敕
勿復辭行成固請帝曰公朕之舊奈何舍朕去邪泫然流涕行成
惶恐不得已復視事未幾卒於尚書省舍年六十七詔九品以上
就第哭比斂三遣使賜內衣服尚宮宿其家護視贈開府儀同三
司并州都督祭以少牢謚曰定弘道元年詔配享高宗廟廷族子
易之昌宗
易之幼以門廕仕累遷尚乘奉御既冠頎皙美姿製音技多所曉

《新唐书》对张平高、张行成的记载

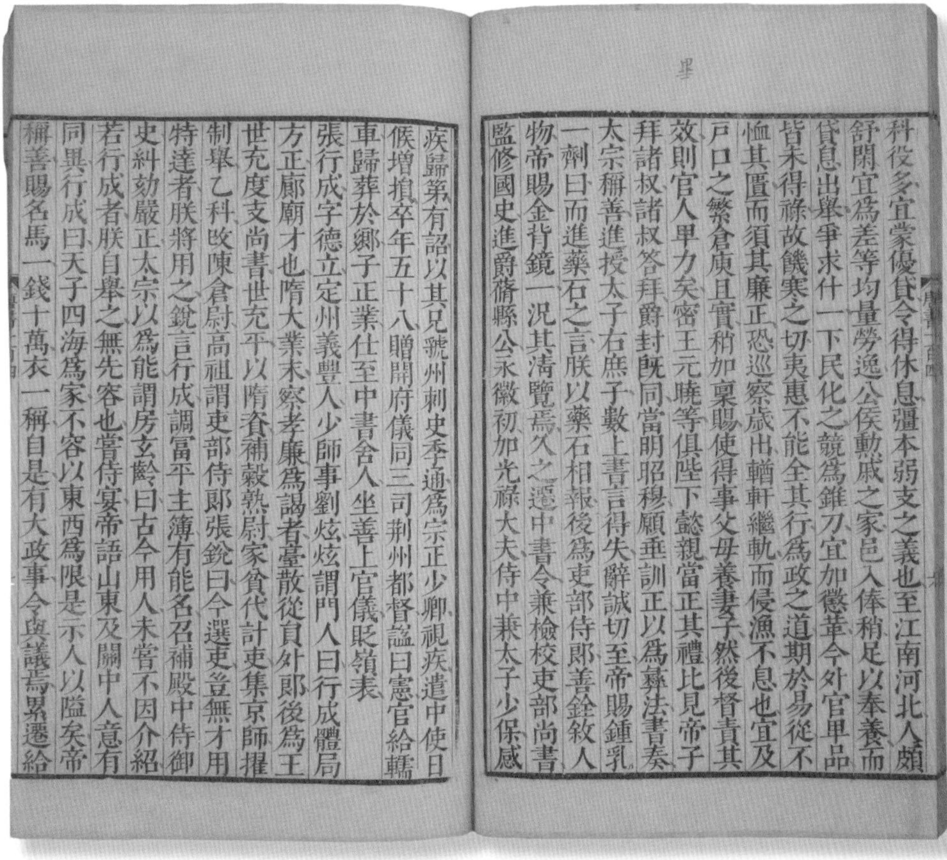

科役多宜蒙優貸令得休息彊本弱支之義也至江南河北人頗
舒閑宜爲差等均勞逸公侯勳戚之家邑體稍足以奉養而
貸息出舉爭求什一下民化之競爲錐刀之政加懲革今外官卑品
皆未得祿故爲饑寒之切夷惠不能全其行爲政之道期於易從不
恤其置而須其廉正恐巡察歲出輶軒繼軌而侵漁不息也宜及
戶口之繁倉廥且實稍加裹贍使得事父母養妻子然後督責其
效則官人甲力矢密王元曉等俱陛下誴親當正其禮比見帝子
拜諸叔叔答拜爵封旣同當明昭穆顧垂訓正以爲舉法書奏
太宗稱善進授太子右庶子數上書言得失辟誠至帝賜鍾乳
一劑曰而進藥石之言以藥石相報後爲吏部侍郎善銓敘人
物帝賜金背鏡一況其清覽爲久之遷中書令兼檢校吏部尚書
監修國史進爵猗縣公永徽初加光祿大夫侍中兼太子少保感

疾歸第有詔以其見虢州刺史委通爲宗正少卿視疾遣中使曰
候增損卒年五十八贈開府儀同三司荊州都督諡曰憲官給轜
車歸葬於鄉子正業仕至中書舍人坐善上官儀貶領表
張行成字德立定州義豐人少師事劉炫炫謂門人曰行成體局
方正廊廟才也隋大業末察孝廉爲謁者臺散從員外郎後爲王
世充度支尚書世充平以隋資補穀熟尉家貧代計集京師
制舉乙科改陳倉尉高祖謂吏部侍郎張銳曰今選吏登科無才用
特達者朕將用之銳言行成調富平主簿有能名補殿中侍御
史科劾嚴正太宗以爲能謂房玄齡曰古今用人未嘗不因介紹
若行成者朕自舉之無先容也嘗侍宴帝語山東及關中人意有
同異行成曰天子四海爲家不容以東西爲限是不以臨奏帝
稱善賜名馬一錢十萬衣一稱目是有大政事令與議焉累遷給

# 六．
# 唐关播宅
# 白居易东亭
位置 │ 西十三舍前

## Site of Residence of Guan Bo of the Tang Dynasty
## Bai Juyi Eastern Pavilion
Front to West Dorm Building 13

窗前有竹玩，门外有酒酤。何以待君子，数竿对一壶。

关播（719—797 年），字务元，卫州汲县（今河南卫辉）人，唐朝宰相，汉寿亭侯关羽之后。旧宅位于常乐坊。

白居易（772—846 年），字乐天，号香山居士，又号醉吟先生，祖籍太原，生于河南新郑，唐代著名诗人。贞元十九年（803 年），白居易通过书判拔萃科考试，授秘书省校书郎，赁常乐坊关播旧宅东亭居住。从贞元十九年春到元和元年（806 年）初，白居易在此居住约三年时间。白居易在常乐坊东亭的生活，可以从他的一首诗《常乐里闲居偶题十六韵兼寄刘十五公》中窥见：帝都名利场，鸡鸣无安居。独有懒慢者，日高头未梳……从诗里看出来，白居易在东亭的生活很是清闲，偶尔还会和朋友相聚同游。

住进东亭的第二天，白居易散步发现此处有一丛竹，得知是关相国亲手所植。白居易进行了一番修剪，并作《养竹记》书于亭壁。文曰：竹似贤，何哉？竹本固，固以树德，君子见其本，则思善建不拔者。竹性直，直以立身，君子见其性，则思中立不倚者。竹心空，空以体道，君子见其心，则思应用虚受者。竹节贞，贞以立志，君子见其节，则思砥砺名行，夷险一致者。夫如是，故君子人多树之为庭焉。

这篇《养竹记》后由著名学者霍松林先生手书，刻在西安交大校园东 7 宿舍前白居易塑像旁。竹的根系牢固，秉性刚直，虚心而坚贞，这是白居易所追求的君子品格，不也正是对后来莘莘学子修身养性的箴言隽语吗？

Guan Bo (719 - 797) served as the top official in the Tang dynasty and his old residence was located in Changle Fang.

Bai Juyi (772 - 846) was a renowned poet of the Tang dynasty, also known as the Hermit of Xiangshan. In 803, Bai passed the imperial examination and the selective examination of the official department and served as an official in Imperial Library. During that period, he lived at Eastern Pavilion of Guan Bo's old residence, which he rented for almost three years from the spring of 803 to early 806.

It seemed that Bai lived a quite leisure life at Eastern Pavilion of Changle Square and he gathered with his friends occasionally, which was reflected in his poems. One of his famous poems created there was *Planting Bamboos*. Bamboos are deep rooted, straight, firm and hollow, which in Chinese culture denote resoluteness, straightness, honor, modesty and faithfulness. These were exactly Bai Juyi's pursuit of a gentleman's character, which can also be what the young students develop themselves into.

The manuscript of *Planting Bamboos* by the famous scholar Huo Songlin was engraved on a wall beside the Statue of Bai Juyi at the front side to the East Dorm Building 7 of Xi 'an Jiaotong University.

梁郡陳夫人閒夷路由於斯當建中貞元之
際兵燹其民火然由其邑大田生荊棘於
斯以為政作事者其難乎去春叔父自徐州士曹參軍
遷歙令於邑之令以約己以清白納人以簡直立事以延
毅以清白故官吏不敢侵于民以簡直故獄訟不得盤
于庭以延毅故軍困兵又曰居二年民用康
政用戰乃曰儲蓄邪之木命先當困桑曰公專吏之所
密而次圖廳事取於土物取村於農素
然後曹約甚其力績徙稱其位儉而至醌壯不至驕死
斯無庶濂之憂視事有朝夕之利官由是而立政由是
無庶濂之憂

而舉民由是而又建一物而三事成其孰不聽之哉鳴
呼吾家世以清簡密為貽燕之訓叔父奉而行之不敢
失墜小子舉而書之亦無愧鮮春其官邑之
之有亡田賦而上下益存乎圓課此省置風物
新聽之時制與叔父作為之所由也先是邑先書令但記
壁無紀則賢姓字湮泯無聞之而今而後居名厥位者經
其年月名氏自叔父始時貞元十九年冬十月一日記

養竹記

竹似賢何哉竹本固固以樹德君子見其本則思善建
不拔者竹性直直以立身君子見其性則思中立不倚

者竹心空空以體道君子見其心則思應用虛受者竹
節貞貞以立志君子見其節則思砥礪名行夷險一致
者夫如是故君子人多樹之為庭實焉貞元十九年春
居易以拔萃選及第授校書郎始於長安求假居處得
常樂里故關相國私第之東亭而處之明日履及於亭
之東南隅見叢竹於斯枝葉殄瘁無聲無色詢於關氏
之老則曰此相國之手植者自相國捐館他人假居由
是筐篚者斬焉彗帚者刈焉刑餘之材長無尋焉數無
百焉又有凡草木雜生其中菶茸薈鬱有無竹之
心焉居易又惜其嘗經長者之手而見賤俗人之目且

若是本性猶存乃芟翳薈除糞壤疏其間封其下不終
日而畢於是日出有清陰風來有清聲依依然欣欣然
若有情於感遇也嗟乎竹植物也於人何有哉以其有
似於賢而人愛惜之封植之況其真賢者乎然則竹之
於草木猶賢之於眾庶嗚呼竹不能自異惟人異之賢
不能自異惟用賢者異之故作養竹記書於亭之壁以
貽其後之居斯者亦欲以聞於今之用賢者云

記畫

張氏子得天之和心之衡積為行發為藝九者其畫尤
歟盡無常工以似為工學無常師以真為師故其措

# 七.
# 隋赵景公寺
位置 | 西十舍

# Zhao Jinggong Temple of the Sui Dynasty
West Dorm Building 10

独孤侧帽倾士女，正平摇笔凌王侯。

赵景公寺，为隋文帝独孤皇后为其父独孤信祈福而建。

独孤信（502—557 年），云中郡盛乐城（今内蒙古和林格尔县）人，鲜卑族，西魏、北周时期名将，历任北周骠骑大将军、大司马等职，是西魏八柱国之一。独孤信堪称"史上最强岳父"，三个女儿均为各代皇后。长女是北周明帝宇文毓皇后，谥号明敬；第四女是唐高祖李渊之母，追封元贞；第七女是隋文帝杨坚皇后，谥号文献。说到独孤信，还有一个和他有关的成语，那就是"侧帽风流"。独孤信在秦州时，有一天因打猎到天晚，骑马入城的时候，他的帽子歪了一点。结果到了第二天，很多官吏和百姓都学着独孤信的样子把帽子歪戴着。可见独孤信的潇洒风神和备受爱戴了。

隋朝建立后，以文献皇后父故，独孤信被追封太师、赵国公，谥号为景，因此被称作"赵景公"。该寺建于隋开皇三年（583 年），本名弘善寺，开皇十八年（598 年）改名赵景公寺。寺中壁上曾有唐代画圣吴道子所画《地狱变相图》。

据《酉阳杂俎》记载，赵景公寺前街有一口八角井。这口井与渭河相通。唐宪宗元和年间（806—820 年），有一位公主让她的婢女用银碗到井中取水，不小心银碗掉进了井中。过了几天，有人竟然在渭河中捡到了这只银碗。

Zhao Jinggong Temple was built by Empress Dugu, wife of Emperor Wen of Sui, for the purpose of memorizing her father, Dugu Xin.

Dugu Xin (502 - 557), one of Xianbei nationality, was born in Shengle City of Yunzhong Prefecture (modern Horinger in Inner Mongolia). He was a famous general in the Western Wei and Northern Zhou dynasties and served several important positions.

Dugu Xin could be called "the most powerful father-in-law in history", since his three daughters were all empresses in three dynasties. The oldest daughter was Empress Ming Jing of Northern Zhou dynasty, wife of Emperor of Ming in Northern Zhou. The forth daughter was Empress Yuan Zhen, Emperor Gaozu of Tang (Li Yuan)'s mother. The seventh daughter was Empress Wen Xian, wife of Yang Jian, Emperor Wen of the Sui dynasty.

After the establishment of the Sui dynasty, Dugu Xin, father of Empress Wen Xian, was honored as Grand Master and Duke of Zhao, with posthumous name of Jing after his death, hence being called "Zhao Jinggong". The temple was built in the third year of Kai Huang of the Sui dynasty with the name of Hongshan Temple and was changed into Zhao Jinggong Temple 15 years later. A famous mural Introduction on *the Painting of the Scenes of the Hells* by Wu Tao-tsu (Wu Daozi), a well known painter of the Tang dynasty, was displayed there.

1997 年 3 月，在我校西南区工地出土大柱础石三个（见图一），镌有唐文宗年号的碑石一通。按历史地理位置，此当为隋大兴城、唐长安城常乐坊西南隅的赵景公寺的建筑遗址。

赵景公寺，本名弘善寺。公元 583 年（隋开皇三年）建，598 年改为赵景公寺。该寺座北朝南，门外有眼八角井，门内有画圣吴道子的壁

画，画中鬼神、天王，其笔迹如铁，窃眸欲语，有呼之欲出之感。正如唐诗段成式的《游长安诸寺·常乐坊赵景公寺·吴画联句》所描述的那样：

> 惨淡十堵内，吴生纵狂迹。
> 风云将逼人，神鬼如脱壁。

这所寺院是隋文帝文献皇后为纪念其父独孤信所立。独孤信（503—557），西魏大司马、八柱国，是一位鲜卑族的上层人物。他生有六子七女，长女是北周明敬皇后，四女（追封）为唐元贞皇后（即唐高祖李渊之母），七女乃隋文帝杨坚文献皇后独孤伽罗。三朝国丈，自古少有，故史学界称其为中国第一老丈人。

隋文帝登基后，追封岳父独孤信为赵国公，邑一万户，谥曰景，故其寺院亦名赵景公寺，为隋唐时期著名的皇家寺院。

1981 年，陕西旬阳县城东门外出土一煤精组印，现被陕西历史博物馆收藏，并在魏晋南北朝展室陈列。这枚印章由二十六个多面体组成，印高 4.5 厘米，重 7.75 克（见图二）。它的主人便是中国历史上显赫一时的第一老丈人独孤信。

陕西历史博物馆陈列着中国第一老丈人印章，西安交大校园出土了老丈人寺院的建筑遗迹，这是历史风云人物留给三秦大地的文化遗产，也是留给我们交大人的重要遗产，我们为有此西迁宝地而自豪，并以此文献给交大西迁五十一周年。

校园观通

# 八.
# 明代秦王府宦官墓葬群
## 位置 | 思源学生活动中心西南侧

## Complex of Tombs
## of Eunuchs of the Ming Dynasty
Southwest to Siyuan Student Activities Center

内宦秦王俱泯灭，当时墓志有哀词。

1996 年至 1998 年，西安交通大学在宪梓堂与思源学生活动中心建筑工地先后发现明代墓葬十座。根据出土的七方墓志可知，墓主为明代秦王府宦官。

公元 1368 年，明太祖朱元璋建立明朝后分封诸王，封次子朱樉为秦王，镇守西安。

朱樉（1356—1395 年），明太祖朱元璋第二子，母亲为马皇后，明代九大塞王之一。洪武三年（1370 年），朱樉被封秦王。洪武十一年（1378 年），正式就藩于西安。

交大出土的七方墓志的墓主，是效力于秦王府的宦官，分别是典宝张义、门副施琮、典宝正任伦、门副魏浚、承奉副邹厚、门官郭镐、中官杨春。七方墓志现藏西安交通大学博物馆。

From 1996 to 1998, 10 tombs of the Ming dynasty were discovered at the construction sites of Xianzi Hall and Siyuan Student Activities Center in Xi'an Jiaotong University. According to the inscription on the seven memorial tablets unearthed from the tombs, the owners of the tombs were eunuchs in the residence of prince of Qin of the Ming dynasty.

Zhu Shuang (1356-1395), the second son of Zhu Yuanzhang, Emperor Ming T'ai-tsu (Ming Taizu), was enfeoffed Prince of Qin to guard Xi'an after the establishment of the Ming dynasty. The seven inscribed memorial tablets are now collected in the Museums of Xi'an Jiaotong University.

上：明秦王府典宝正任伦墓志盖　中：明秦王府门副魏浚墓志盖　下：明秦王府门副施琼墓志盖

# 九.
# 唐来济宅
位置 ｜ 化工楼西

## Site of Residence
## of Lai Ji of the Tang Dynasty
West to Chemical Engineering Building

相看白刃血纷纷，死节从来岂顾勋。

来济（610—662年），扬州江都（今江苏扬州）人，隋朝左翊卫大将军来护儿之子。在隋末宇文化叛乱中，来护儿全家被杀，只有来济与弟弟幸免于难。来济虽家境孤苦无依，但发奋上进，唐高祖武德年间（618—626年），获进士出身。贞观十七年（643年）来济任通事舍人，累迁中书舍人，与令狐德棻登一同编撰《晋书》；永徽二年（651年），拜中书侍郎，兼弘文馆学士，监修国史；永徽四年（653年），加同中书门下三品；永徽六年（655年），迁中书令、检校吏部尚书，为正三品文官。

来济为官正直，时常抓住时机劝谏帝王，为百姓陈情。高宗欲废王皇后，立昭仪武则天为"宸妃"。来济等人为江山社稷考虑，表示强烈反对，被武则天所忌恨，显庆二年（657年），因诬奏累贬庭州刺史。龙朔二年（662年），西突厥入寇庭州，来济以必死之心力战阵亡，时年五十三岁。

Lai Ji's (610 - 662) parents died when he was very young, yet he was particularly studious in spite of his helplessness. After passing the palace examination and attaining the degree title of Jinshi, Lai Ji started his career as an official. He was engaged in compiling *The Book the of Jin Dynasty* and was in charge of supervising the revision of imperial history. Later he was promoted to higher position, serving as Minister of Civil Service. Lai Ji was righteous, and often seized the opportunity to admonish Emperor Gaozong and spoke for the people. He made all his efforts to serve the Emperor and the imperial court.

良貶潭州都督明年瑗上言遂良受先帝顧託一德無二向日論
事至誠懇切詎肯令陛下後堯舜而塵史冊哉遭厚謗醜言損陛
下之明折志士之銳況被遷以來再離寒暑其責寬無辜有
以順眾帝曰遂良之情朕知之矣其字戾好犯上朕責之詎有
過邪瑗曰遂良社稷臣蒼蠅點白傅致有罪昔微子去股以亡
張華不死晉不及亂陛下富有四海安於清泰或驅逐舊臣遂不
省察乎帝愈不聽瑗憤自表歸田里不報顯慶二年許敬宗李
義府奏瑗以桂州授遂良桂地倚用武地倚為之謀於是貶振州刺
史瑗奏瑗无忌死義府等復奏瑗與通謀遣使即
殺之既至瑗已死發棺驗視乃還追削官爵籍其家子孫
官奴神龍初武后遺詔復官爵自瑗暨遂良相繼死內外以言為
諱將二十年帝造秦天官御史李善感始上疏極言時人喜之謂

為鳳鳴朝陽

來濟揚州江都人父護兒隋左翊衛大將軍宇文化及難闔門死
之濟幼得免轉側流離而篤志為文章善議論曉暢時務擢進士
貞觀中累遷通事舍人太子承乾敗太宗問何以處之莫敢
對濟曰陛下上不失為慈父太子得盡天年則善帝納之除考功
員外郎十八年初置太子司議郎高其選而以濟為之兼崇賢館
直學士遷中書舍人永徽二年拜中書侍郎兼弘文館學士監修
國史俄同中書門下三品封南陽縣男遷中書令檢校吏部尚書
帝將以武氏為后宜擇禮義
名家幽閑令淑副四海之望稱神祇之意故文王興姒關雎之
化蒙被百姓其福如彼成帝縱欲以婢為后其禍如此
惟陛下詳察初武氏被寵帝特號宸妃濟與韓瑗諫妃有常員令

# 十.
# 唐郭敬之宅
位置 | 化工楼东

## Site of Residence
## of Guo Jingzhi of the Tang Dynasty
East to Chemical Engineering Building

神灵汉代中兴主，功业汾阳异姓王。

郭敬之（667—744 年），唐代著名将领汾阳王郭子仪之父。郭敬之家族源于太原郭氏。自西汉冯翊太守郭孟儒开始，郭氏世居太原，后有一支迁徙到华州郑县（今陕西渭南市华州区），郭敬之就是这一支的后代。

据郭子仪为其父所立、颜真卿撰书的《郭公庙碑铭》，郭敬之相貌不凡："身长八尺二寸，行中絜矩，声如洪钟，河目电照，虬须猬磔。进退闲雅，望之若神。"郭敬之初任涪州录事参军，累迁绥、渭、桂、寿、泗五州刺史，退休后回到京城长安常乐坊，安度晚年，唐玄宗天宝三年（744 年）因病去世，享年七十八岁。

说到郭敬之，不得不提的当然就是他的次子、一代名将郭子仪了。郭子仪戎马一生，功勋卓著。在平定安史之乱的战争中，郭子仪指挥了攻克河北诸郡之战、收复两京之战、邺城之战等重大战役。安史之乱后，郭子仪又设计抵御吐蕃，二次收复长安，并威服叛将，平定河东。《旧唐书》称他"再造王室，勋高一代"，为稳固李唐江山立下了汗马功劳。

因为郭子仪日渐显贵，郭敬之在死后也备受尊荣。唐肃宗乾元元年（758 年）二月，唐肃宗下诏追赠郭敬之为太保、祁国公。唐代宗广德二年（764 年），郭子仪为其父树立一块家庙碑，即《郭公庙碑铭》。此碑由唐代宗李豫亲自隶书题额。此碑道劲浑厚，疏朗流畅，如银钩铁画，龙跳虎卧，据后代学者考证，碑文系由著名书法家颜真卿撰书。该碑原树立于陕西西安府布政司署内，1950 年移入西安碑林。

Guo Jingzhi (667 - 744), an official of the Tang dynasty, was described in inscription on the Stele of Guogong Temple as an extraordinary man in terms of his appearance and temperament. He served as governor at five prefectures supervising and reporting illegal acts to the court. After retirement, he returned to the capital Chang'an, living at Changle Square.

Guo Ziyi, Guo Jingzhi's second son, was a famous general and made outstanding contributions in his military life. He played a critical role in pacifying the An-Shi Rebellion and fighting against Tubo, which was incomparable in stabilizing the Tang dynasty. Guo Jingzhi, as a result, was honored as Duke of Qi after his death and Guo Ziyi built a stele for his father known as *the Stele of Guogong Temple*. According to archaeological studies, the inscription was written by Yan Zhenqing, a renowned calligrapher of the Tang dynasty. The stele was originally placed in Administrative Divisions of Xi'an Prefecture, Shaanxi Province (the old residence of Guo Ziyi), and was collected in The Stele Forest Museum in Xi'an in 1950.

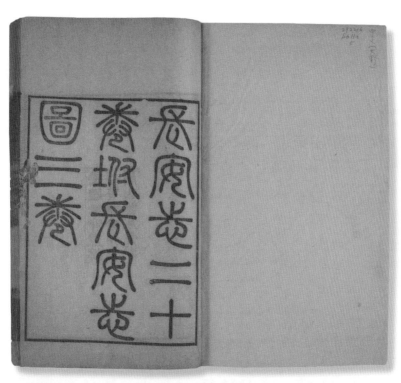

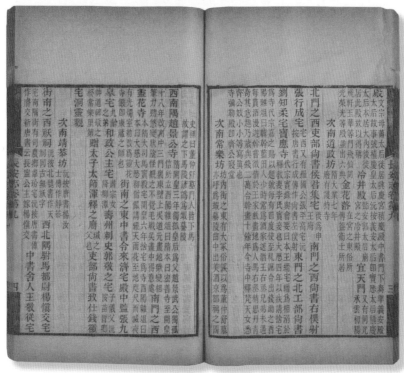

# 十一.
# 唐和政公主宅
位置 ｜ 南洋学术交流中心前

## Site of Residence
## of Princess Hezheng of the Tang Dynasty
Front to Nanyang Academic Exchange Center

主家别墅帝城隈，无劳海上觅蓬莱。

和政公主（729—764 年），唐肃宗李亨第三女，生母为章敬皇后吴氏。和政公主是唐代宗同母妹，以敏惠纯孝著称。章敬皇后早逝，三岁的和政公主由韦妃抚养，对韦妃恭敬孝顺有如生母。她多才多艺，深受肃宗喜欢，后下嫁河东柳潭为妻，育有五子三女。

安史之乱时，唐玄宗和皇室大臣一同南逃，和政公主主动把马让给寡居的二姐宁国公主，自己和丈夫在山野间徒步行走，烧火做饭，侍奉宁国公主。安史之乱后，朝廷凋敝，和政公主把家产捐献给王公戚属，自己缝补衣裳，生活俭朴。和政公主还十分关心政事和民间疾苦。代宗广德二年（764 年），吐蕃入侵，公主入宫觐见代宗，出谋划策。时值六月，天气炎热，和政公主积劳成疾，第二天不幸病逝，年仅三十六岁。

Princess Hezheng (729 - 764) was the third daughter of Li Heng, Emperor Suzong of Tang. She was intelligent, kind, plain, filial and versatile, so she was deeply adored by the Emperor. The princess was also very concerned with the state affairs and the sufferings of the ordinary people. When An Shi Rebellion took place, Emperor Xuanzong and court officials to the north, and Princess Hezheng let Princess Ningguo in widowhood ride her horse, and she and her husband walked along and cooked for her. After the An Shi Rebellion, the empire was destitute, and Princess Hezheng donated her properties to the members of the imperial family, and she herself patched clothes and lived a simple life. Because she was also concerned with state affairs, when Tufan invaded the Tang empire in the year 764, she talked with Daizong then, and made plabns to defeat the invasion. It was hot summer, and she was working so hard as to catch a disease, and died quickly at the age of 36.

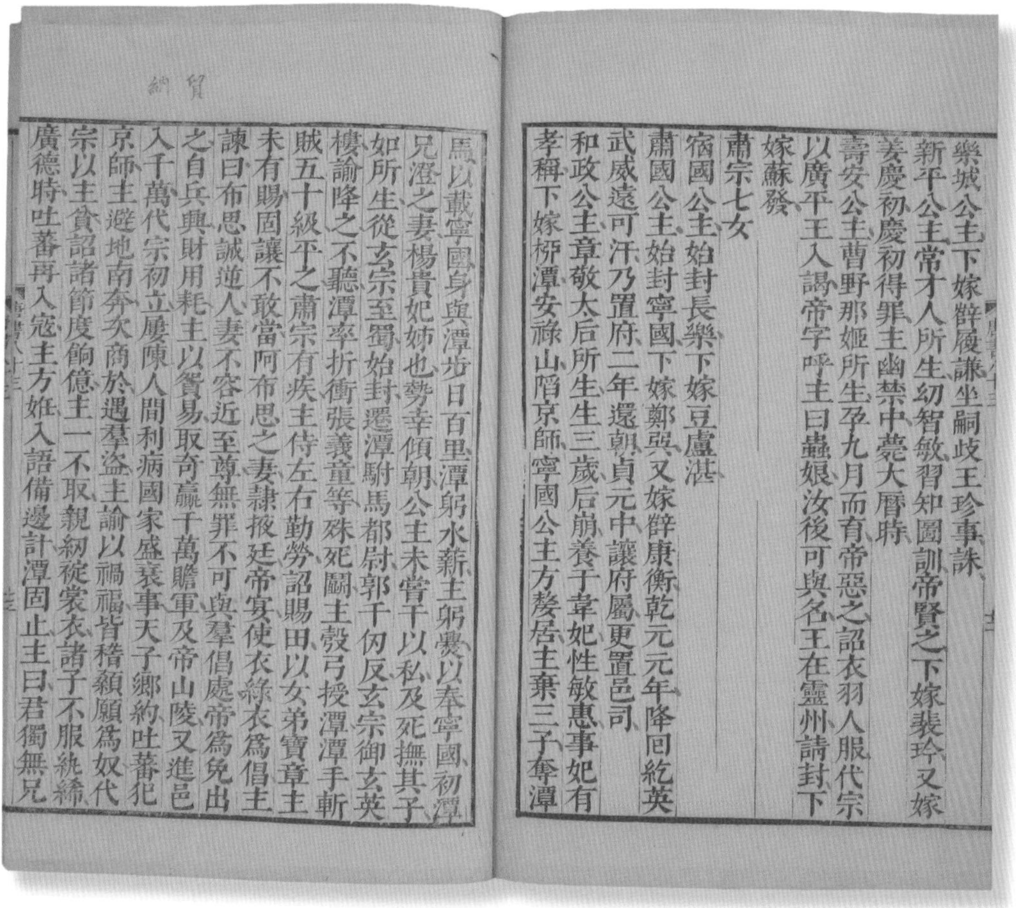

榮城公主下嫁辭屢謙坐嗣歧王珍事誅
新平公主常才人所生初智敬習知圖讖帝賢之下嫁裴玲又嫁
美慶初慶初得罪主幽禁中薨大曆晴
壽安公主曹野那姬所生孕九月而育帝惡之詔衣羽人服代宗
以廣平王入謁帝字呼主曰蟲娘汝後可與名王在靈州請封下
嫁蘇發
肅宗七女
宿國公主始封長樂下嫁豆盧湛
蕭國公主始封寧國下嫁鄭巽與又嫁辭康衡乾元元年降回紇英
武威遠可汗乃置府二年還朝貞元中讓府屬更置邑司
和政公主章敬太后所生生三歲崩養子韋妃性敏惠事妃有
孝稱下嫁柳潭安祿山陷京師寧國公主方娶居主兼三子奉潭

馬以載寧國身與潭步日百里潭躬水薪主躬襲以奉寧國初潭
兄澄之妻楊貴妃姊也勢幸傾朝公主未嘗干以私及死撫其子
如所生從玄宗至蜀始封遷潭駙馬都尉郭千仞反玄宗御玄英
樓諭降之不聽潭率折衝張義童等殊死鬬主彀弓授潭手斬
賊五十級平之肅宗有疾主侍左右勤勞詔賜田以女弟寶章主
未有賜固讓不敢當阿布思之妻隸掖廷帝宴使衣綾衣爲倡主
諫曰布思誠逆人妻不容近至尊無罪不可與羣倡處帝爲免出
之自兵興財用耗主以賓易取奇贏千萬贍軍及帝鄉陵又進邑
入千萬代宗初立屢陳人間利病國家盛衰事天子鄉約吐蕃犯
京師主避地南奔災商於遇羣盜主諭以禍福皆願爲奴代
宗以主貧詔諸節度餉億主一不取親紉裳衣者子不服紈綺
廣德特吐蕃再入寇主方姓入語備邊計潭固止主曰君獨無兄

# 十二．
# 唐虾蟆陵旧址
位置 | 东十一、十二舍之间

## Site of Hamaling of the Tang Dynasty
Between East Dorm Buildings 11 and 12

自言本是京城女，家在虾蟆陵下住。

白居易《琵琶行》里提到的琵琶女所居住的"虾蟆陵"在哪里？根据北宋宋敏求在《长安志》卷九"常乐坊"条中的记述："坊内街之东有大冢，俗误以为董仲舒墓，亦呼为虾蟆陵。"《长安志》卷十一"万年县"下又记载："虾蟆陵在县南六里。"而根据文献图志记载，唐京兆府万年县治所在的位置就在今天西安市和平门西北的东县门街一带，距交大距离正为六里。元代李好文《长安志图》"城南名胜古迹图"中，长安城南有"烟（胭）脂坡"，胭脂坡旁也标注有"下马陵"三字，并注曰："本董仲舒墓。过者皆下马，故名，讹为虾蟆陵。"

1998年10月25日，交大校园在修建浴池时，发现一座汉代砖砌古墓，很有可能是董仲舒之墓，即后代所说的"虾蟆陵"。琵琶女所居住的"虾蟆陵"在唐代长安为歌伎集中之地。而且此地又盛产美酒。据清代徐松的《唐两京城坊考》，坊中"曲中出美酒，京都称之"。因此，大诗人白居易的故居东亭的旁边，竟然就是他在江州所遇到的琵琶女的故居，两人可真是有缘分了。

Hamaling was mentioned in Bai Juyi's *Song of the Lute Player* and do you know where it is? According to historical records, Hamaling was at Changle Square, near today's East Xianmen Street, northwest of Heping Gate in Xi'an about three kilometers away from Xi'an Jiaotong University. In *Chang'an Zhi Tu* by Li Haowen of the Yuan dynasty, it writes: "There is 'Yanzhipo' Slope in the south of Chang'an City, beside which three Chinese characters "Xia Ma Ling" are marked with a note: Dong Zhongshu's (an accomplished philosopher, educator, and politician of the Western Han Dynasty) tomb is here so those who have passed all get off their horses (Xia Ma in Chinese) to show respect for him. "Xia Ma" reads "Ha Ma" by local accent, hence the name Hamaling.

On October 25, 1998, a brick tomb of the Han dynasty was discovered in Xi'an Jiaotong University at the construction site of a bath house, which was said to be the Dong Zhongshu's tomb, also "Hamaling" as called by later generations.

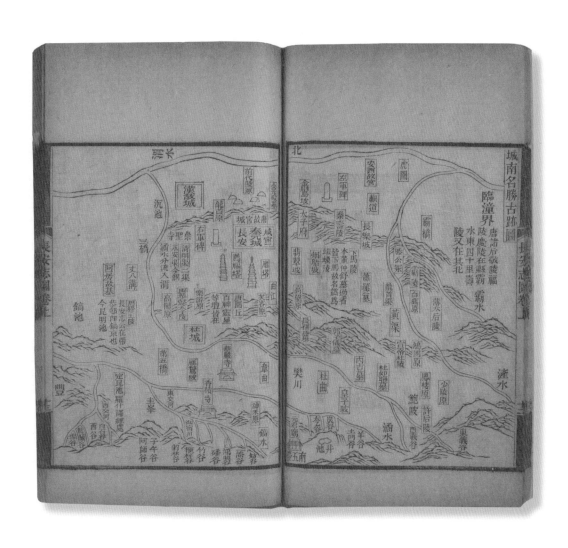

《长安志图》中的"城南名胜古迹图"

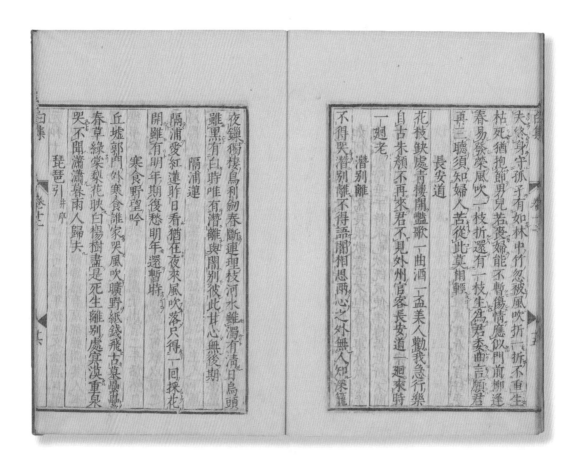

夫終身守孤子看如林中竹忽被風吹折一折不重生
枯死猶抱節男兒若喪婦能不暫傷情應似門前柳逢
春易發榮風吹一枝折還有一枝生為君委曲言願君
再三聽須知婦人苦從此莫相輕

長安道
花枝缺處青樓開豔歌一曲酒一盃美人勸我急行樂
自古朱顏不再來君不見外州官客長安道一迴來時
一迴老

潛別離
不得哭潛別離不得語闇相思兩心之外無人知深籠

夜鎖雙棲鳥利劍春斷連理枝河水雖濁有清日烏頭
雖黑有白時唯有潛離與闇別彼此甚心無後期

隔浦蓮
隔浦愛紅蓮昨日看猶在夜來風吹落只得一迴採花
開離有明年期後愁明年還暫時

寒食野望吟
丘墟郭門外寒食誰家哭風吹曠野紙錢飛古墓纍纍
春草綠棠梨花映白楊樹盡是死生離別處冥寞重泉
哭不聞蕭蕭暮雨人歸去

琵琶引并序

元和十年予左遷九江郡司馬明年秋送客湓浦口聞舟中夜彈琵琶者聽其音錚錚然有京都聲問其人本長安倡女嘗學琵琶於穆曹二善才年長色衰委身為賈人婦遂命酒使快彈數曲曲罷憫然自敘少小時歡樂事今漂淪憔悴轉徙於江湖間予出官二年恬然自安感斯人言是夕始覺有遷謫意因為長句歌以贈之凡六百一十二言命曰琵琶行

潯陽江頭夜送客楓葉荻花秋瑟瑟主人下馬客在船舉酒欲飲無管絃醉不成歡慘將別別時茫茫江浸月忽聞水上琵琶聲主人忘歸客不發尋聲闇問彈者誰

琵琶聲停欲語遲移船相近邀相見添酒回燈重開宴千呼萬喚始出來猶抱琵琶半遮面轉軸撥絃三兩聲未成曲調先有情絃絃掩抑聲聲思似訴平生不得志低眉信手續續彈說盡心中無限事輕攏慢撚抹復挑初為霓裳後六么大絃嘈嘈如急雨小絃切切如私語嘈嘈切切錯雜彈大珠小珠落玉盤間關鶯語花底滑幽咽泉流水下灘水泉冷澀絃凝絕凝絕不通聲暫歇別有幽愁暗恨生此時無聲勝有聲銀瓶乍破水漿迸鐵騎突出刀槍鳴曲終收撥當心畫四絃一聲如裂帛東船西舫悄無言唯見江心秋月白

沉吟放撥插絃中整頓衣裳起斂容自言本是京城女家在蝦蟆陵下住十三學得琵琶成名屬教坊第一部曲罷曾教善才服妝成每被秋娘妒五陵年少爭纏頭一曲紅綃不知數鈿頭銀篦擊節碎血色羅裙翻酒汚今年歡笑復明年秋月春風等閒度弟走從軍阿姨死暮去朝來顏色故門前冷落鞍馬稀老大嫁作商人婦商人重利輕別離前月浮梁買茶去去來江口守空船遶船明月江水寒夜深忽夢少年事夢啼妝淚紅闌干我聞琵琶已嘆息又聞此語重唧唧同是天涯淪落人相逢何必曾相識我從去年辭帝京謫居臥病潯陽城

潯陽地僻無音樂終歲不聞絲竹聲住近湓江地低濕黃蘆苦竹遶宅生其間旦暮聞何物杜鵑啼血猿哀鳴春江花朝秋月夜往往取酒還獨傾豈無山歌與村笛嘔啞嘲哳難為聽今夜聞君琵琶語如聽仙樂耳暫明莫辭更坐彈一曲為君翻作琵琶行感我此言良久立卻坐促絃絃轉急淒淒不似向前聲滿座重聞皆掩泣座中泣下誰最多江州司馬青衫濕

簡簡吟

蘇家小女名簡簡芙蓉花腮柳葉眼十一把鏡學點妝十二抽針能繡裳十三行坐事調品不肯迷頭白地藏

# 十三.
# 唐刘知柔宅
位置 | 交大一村

# Site of Residence
# of Liu Zhirou of the Tang Dynasty
XJTU Residential Compound 1

刘郎才思长卿看，帝里风光驻马鞍。

刘知柔（649—723年），彭城（今江苏徐州）人，为汉宣帝之子楚孝王刘嚣后代。其曾祖为北魏骠骑大将军、北州刺史刘祎，曾祖为北齐散骑常侍、文林馆学士刘岷，祖父为朝散大夫、陈留县长刘元邃，父亲为宋州司马、赠徐州刺史刘藏器。成长于官宦之家的刘知柔还是唐代著名史学家刘知几的兄长。刘知几所著的《史通》是中国乃至世界首部系统性的史学理论专著，具有十分重要的地位和价值。

据碑文记载，刘知柔"仪形硕伟，风神散逸"，而且性格幽静镇重，立志好学，勤俭谦虚（《唐赠太子少保刘知柔神道碑》）。进士出身后，刘知柔历任国子司业、工部尚书、太子宾客等显要职务，后被封为彭城县侯，去世后被追赠太子少保。刘知柔去世后，由著名文学家李邕撰写神道碑，即《唐赠太子少保刘知柔神道碑》，收录于《全唐文》中。

Liu Zhirou (649 - 723) was born and grew up in an official family, and he was the elder brother of Liu Zhiji, a famous historian in the Tang dynasty. *Generality of Historiography* by Liu Zhiji was the first systematic monograph on historical theory in China and even in the world, which is of great value in historical studies.

According to the inscription on a tablet in a tomb, Liu Zhirou was handsome, sedate, ambitious, and industrious. After passing the palace examination and attaining the degree title of Jinshi, he successively served at different important positions, including Director of Imperial Academy, Minister of Works, and Governor of Pengcheng County, etc.

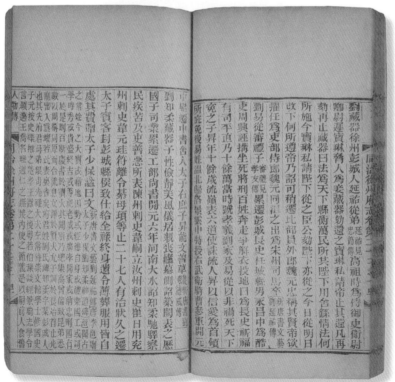

# 十四.
# 唐代六件玉花簪头出土地

位置 ｜ 交大一村

# Excavation Site
# of Six Jade Flower Hairpins of the Tang Dynasty

XJTU Residential Compound 1

手把玉簪敲砌竹，清歌一曲月如霜。

1979年，西安交通大学在建一村西部住宅楼时，出土了六件唐代玉花簪头，也叫"白玉头钗花饰"。簪头整体呈叶片形状，上面雕刻有植物花卉纹饰，雕琢细腻，秀雅富贵，为国家一级文物，现藏于西安博物院。

唐代妇女发髻的样式很复杂，用来束发的首饰种类也很多。单股的为簪，双股的为钗。簪的质地有竹、角、金、银、牙、玉等多种。从样式上来说，有些簪头的图案十分繁缛，除固定发髻的功能之外，还有装饰的功能。

玉簪有一个别名，叫作"玉搔头"。这个名字从何而来呢？据《西京杂记》记载，汉武帝曾经在宠爱的李夫人头上取下玉簪，拿来搔头上的痒，因而玉簪得名"玉搔头"。中唐诗人刘禹锡有一首《和乐天春词》："新妆宜面下朱楼，深锁春光一院愁。行到中庭数花朵，蜻蜓飞上玉搔头。"诗中这位可爱的女子所佩戴的就是玉簪了。

In 1979, six jade flower hairpins of the Tang dynasty were unearthed in Xi'an Jiaotong University when residential buildings were being built in the west of XJTU Residential Compound 1. The six jade hairpins were all in the shape of leaves with exquisite patterns of plants and flowers. The elegant hairpins are national first-class cultural relics and are now collected in Xi'an Museum.

Hairpins were a kind of jewelry for women to fix their hair buns in the Tang dynasty. Hairpins, which can be made of bamboo, horn, gold, silver, jade then and so on, were not only for holding hair, but also for decoration.

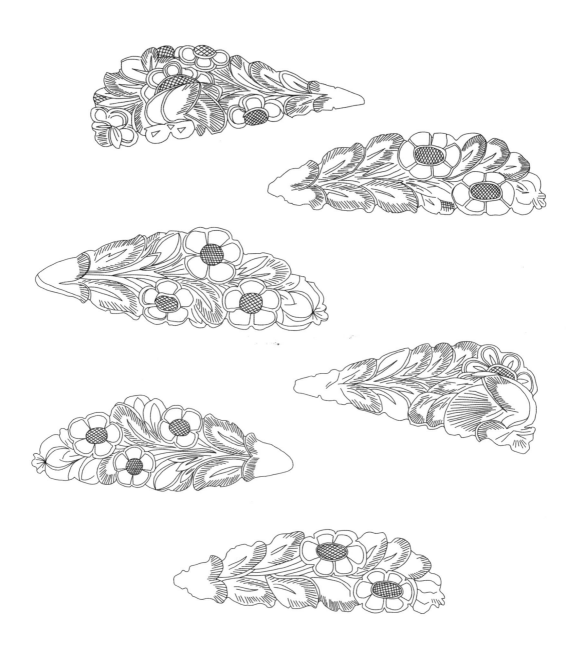

六件玉花簪头（手绘）

# 十五.
## 唐王缙宅
## 宝应寺旧址
位置 │ 交大一村

# Site of Residence of Wang Jin
# Site of Baoying Temple of the Tang Dynasty
XJTU Residential Compound 1

位高汤左相，权总汉诸侯。

王缙（702—781年），字夏卿，祁县（今山西祁县）人。唐代宰相，著名诗人王维之弟。王缙自幼好学，与王维皆以名闻。此外，王缙还是一名书法家，擅长草书和隶书。他科举中第后，授侍御史，历任太原尹、河南副元帅、河东节度使，两拜门下侍郎、同平章事。王缙曾协助李光弼平定安史之乱，后因附和权臣元载，被贬为括州刺史，终迁太子宾客、分司东都，被封为齐国公。王缙于建中二年（781年）去世，时年八十二岁。在安史之乱中，王缙的兄长王维因被安禄山俘虏而接受伪职，因此在叛乱平息后被朝廷审查。王缙请求免除自己的官职来为兄长赎罪，因此使王维得到从宽处理。兄弟情深，可见一斑。

宝应寺原为王缙府第。王维有一首《戏赠张五弟諲三首》，其自注云："时在常乐东园走笔成。"可见，王维也曾经在常乐坊居住，他的居所很有可能就在王缙府中的东园。王氏兄弟的母亲崔氏是佛教徒。王维则是"诗佛"。在他们的影响下，王缙也沉溺佛教。在妻子李氏亡故后，王缙舍宅为寺，寺中的弥勒堂就是王缙的寝室。唐代宗为宝应寺题写了寺名。

宝应寺中曾有唐代著名画家韩干所画的弥勒和菩萨像。据史料记载，韩干是陕西蓝田人，年少时只是一个酒馆的雇工。有一次他到王家来讨酒债，等待的时候就在地上画画。王维看到韩干的戏作，认为他具有绘画天赋，就资助他学画。韩干学成后被召为宫廷画师。皇帝命他拜当时画马名家陈闳为师，韩干不听命，反而搬到马厩里和马夫一起居住。韩干与王氏兄弟的渊源很深，据说宝应寺中壁画上的天女都是依照王缙家中歌伎的样子来画的。

Wang Jin (702 – 781) served as a top official of the Tang dynasty and took many important positions. Wang Jin, talented and studious in his youth, was a calligrapher and was as famous as his elder brother Wang Wei, a famous poet of Tang. He was honored as Duke of Qi and died in 781 at the age of 82.

Baoying Temple used to be Wang Jin's residence at Daozheng Fang. Wang Jin, influenced by his mother, was a devout follower of Buddhism. After the death of his wife, Wang Jin got his residence converted into a temple (Baoying Temple) in order to pray for blessing for his dead wife. The Maitreya Hall in the temple was his bedroom. Emperor Daizong of Tang inscribed the name for the temple.

王维《辋川图》

# 十六.
# 西汉壁画墓发掘地
位置 ｜ 交大二村

# Excavation Site
# of a Western Han Dynasty Tomb with Murals
XJTU Residential Compound 2

古墓丹青世所无，当年萧傅亦堪吁。

1987 年 4 月，西安交通大学附属小学建造教学楼时，在挖掘地基时发现了一座西汉壁画墓，这是迄今为止全国发现的九座西汉壁画墓之一，被考古界命名为"西安交通大学西汉壁画墓"。

墓内主室顶部及四壁，密集地填绘着各种图案，没有丝毫空白，绘有祥云缭绕、仙鹤飞翔、瑞兽遍布，且包含相对完整的我国古代天文中的"二十八星宿"图。这幅壁画在天文、美术等领域均具有相当高的价值。

据考证，此墓有可能为汉宣帝时期太子太傅萧望之之墓。萧望之（公元前 106—前 47 年），字长倩，东海兰陵（今山东苍山）人，后徙杜陵（今陕西西安东南），萧何七世孙。萧望之博通经术，汉宣帝时期历任大鸿胪、太傅等官。汉宣帝甘露三年（公元前 51 年），在长安未央宫北的石渠阁，召开了一次宣帝亲临的学术会议，研讨五经异同。这次中国学术史上著名的"石渠阁会议"即由萧望之主持评议。汉元帝即位后，萧望之辅佐朝政，甚受尊重，后遭宦官弘恭、石显等诬告下狱，愤而自杀。

In April 1987, a mural tomb of the Western Han dynasty was excavated in XJTU-affiliated Primary School during construction of teaching buildings. It was one of the nine mural tombs of the Western Han dynasty hitherto excavated in China and was named as "XJTU mural tomb of the western Han dynasty".

The top and four walls of the main chamber in the tomb were densely painted with various patterns such as auspicious clouds, the flying white cranes, and propitious animals, and relatively complete pictures of "Twenty-eight Constellations" in ancient Chinese astronomy. This mural is of very high value in astronomy, fine arts and other fields.

According to archaeological studies, the owner of the tome was probably Xiao Wangzhi, the Grand Tutor of the Crown Prince (Taifu) during the reign of Emperor Xuan of Han. Xiao Wangzhi ( 106BC-47BC) was proficient in The Five Classics of Confucianism, and took several important positions. In 51 BC during the reign of Emperor Xuan of Han, Xiao hosted an academic conference 51BC in Shiqu Pavilion, north to Weiyang Palace, with the presence of the Emperor to discuss the distinctions and similarities among The Five Classics, which was known as "Shiqu Pavilion Conference" in Chinese academic history. After Emperor Yuan of Han ascended the throne, Xiao assisted him and was highly respected.

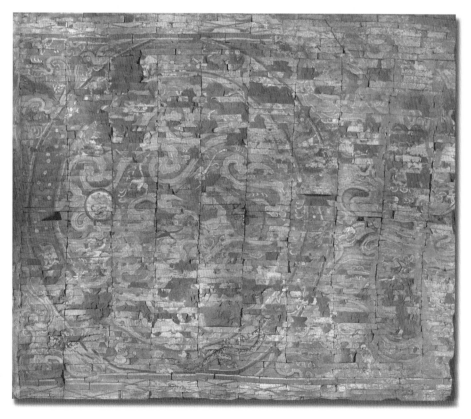

左上：西安交大西汉墓出土的长生未央砖　右上："二十八星宿"图（手绘）　下：西安交大西汉主墓室拱形顶部壁画

# 十七.
# 唐安禄山宅

## Site of Residence
## of An Lushan of the Tang Dynasty

胡兵忽从天上来，逆胡亦是好雄才。

　　唐代韦述《两京记》中载："禄山旧宅在道政坊，玄宗以起隘陋，更于亲仁坊选宽爽之地，出内库钱，更造宅焉。"可见唐代著名藩镇叛臣、安史之乱的发动者安禄山也曾经在道政坊居住。

　　安禄山（703—757年），本姓康，字轧荦山，营州柳城（今辽宁朝阳）人，粟特族。安禄山出身西域康国，其父早逝。其母阿史德氏，改嫁安延偃，从此他改为安姓。安禄山精通九番语言，早年骁勇善战，屡建功勋。天宝年间（742—756年），安禄山备受唐玄宗宠信，平步青云，兼任平卢、范阳和河东三镇节度使。天宝十四年（755年），安禄山以诛杀宰相杨国忠为名，发动安史之乱，一路攻破两京，建立伪燕政权。至德二年（757年），安禄山为嫡次子安庆绪指使宦官所害。

According to historical records, An Lushan used to live in Daozheng Fang. An Lushan (703 - 757), proficient in several languages, was valiant and good at fighting, so he created meritorious military services and made great contributions at his early age. That was why An Lushan was extremely favored by Emperor Xuanzong, which made him rise through the ranks of being the military governor of three townships. In 755, An Lushan launched a rebellion (the so called AnShi Rebellion) which was quelled eventually.

# 十八.
# 唐韩翃与柳氏重会处

# Reunion Site
# for Tang Dynasty's Han Hong and His Wife

章台柳，章台柳，昔日青青今在否？

《柳氏传》是中唐文人许尧佐的著名传奇小说，其中所写的韩翃与柳氏的爱情故事一向为人津津乐道。而这个爱情故事设定的重要地理位置就是唐代长安城的道政坊。

韩翃是唐代天宝年间很有名的诗人，但十分贫寒困顿。他的好朋友李生是一个爱惜人才的豪迈之士，让韩翃住在他家里。李生有一个漂亮的宠姬柳氏，容貌美丽，又喜欢吟诗作赋。柳氏偷偷从门里看韩翃，觉得他气质不凡，对韩翃十分倾心。李生知道后，就将柳氏配于韩翃。二人郎才女貌，十分幸福。第二年，韩翃中了进士，柳氏劝他回老家报喜探亲。谁知韩翃离开之后，安史之乱爆发，叛军攻占了长安。有一个叫沙吒利的蕃将，仗着自己在平叛中有战功，将柳氏强夺。韩翃回长安后，一日偶遇柳氏车驾，相约次日在道政坊门口相会。在道政坊前，柳氏十分绝望，赠给韩翃一个玉盒，跟他道别，就回车而去。韩翃悲痛不已。这时候，正好韩翃当年的同僚正在附近的酒楼喝酒，让人来请韩翃一起赴约。韩翃勉强答应，在席间却失魂落魄。一个叫许俊的虞候得知事情原委，便马上骑马前往沙吒利的府第，等沙吒利出门走远，就进门大呼将军得了急病要来找夫人，奴仆没有敢不听命的。许俊将韩翃的书札给柳氏看后，就将柳氏挟持在马上，又回到了酒楼。沙吒利当然不会罢休。韩翃、许俊就找到了淄青节度使侯希逸。侯希逸起草一纸状书给皇帝，叙述沙吒利抢夺柳氏经过。皇帝看了，下诏将柳氏还给了韩翃。为了平息沙吒利的怨气，皇帝也赏了他二百万钱。柳氏和韩翃从此幸福地生活在了一起。

Han Hong and his wife were two main characters in *The Story of Lady Liu*, a legendary love story by Xu Yaozuo, a man of letter in the Tang dynasty, which happened in Daozheng Fang of Chang'an City.

Han Hong was a very famous poet in the Tianbao period of the Tang dynasty, but he was very poor and distressed. His good friend Li Sheng was a bold and generous person who cherished talentedscholars, so he let Han Hong live in his house. Li Sheng had a beautiful favorite belle named Ms Liu. Ms The Liu was beautiful, and liked to recite poems and write *fu*, a descriptive prose interspersed with verse. When Ms. Liu secretly looked at Han Hong from inside the door, she felt that he had an extraordinary temperament and was very enamoured of Han Hong. After Li Sheng knew this, he belived that they were a perfect match, so he let Ms Liu marry Han Hong, and they then had a happy life.

In the second year, Han Hong passed the highest imperial examinations, and Ms Liu persuaded him to go home to tell his parents about it. Unexpectedly, upon leaving for his hometown, An Shi Rebellion broke out, and the rebels occupied Chang'an city. A general named Shaachili seized Ms Liu.

Han Hong asked help from Hou Xiyi, the provincial governor in charge of cilvil and military affairs of now Shandong. Hou Xiyi wrote a report to the emperor, describing how Shaachili seized Ms Liu through force. After reading the reporrt, the Emperor issued an edict, and asked that Ms Liu go back to Han Hong'. In order to calm Shaachili for grievances, the emperor also rewarded him with two million dollars. Ms Liu and Han Hong then lived a happy life once again.

# 十九 .
# 唐张九皋宅

## Site of Residence
## of Zhang Jiugao of the Tang Dynasty

草木有本心，何求美人折！

张九皋（690—755年），韶州曲江（今广东韶关）人，为唐代著名宰相张九龄之弟，也是元曲作家张养浩（1270—1329年）的二十三世祖。其早年历任南海郡司户参军、赣县令、始安太守等职，张九龄入相后，张九皋改任南康别驾。

张九龄贬谪荆州后，张九皋亦随贬淮安、彭城等地郡守，后历迁襄、广数州刺史，特授银青光禄大夫。其墓志铭记载："以天宝十四载四月二十日，疾薨薨於西京常乐里之私第，春秋六十有六。"可见张九皋晚年曾居住于常乐坊。

Zhang Jiugao (690 – 755) used to live in Changle Square. He was the younger brother of Zhang Jiuling, the well known official of the Tang dynasty. In his early years, he served as an official in charge of household registration at Nanhai County, Magistrate of Ganxian County, Prefecture Chief in Shian, the Assistant Governor in Nankang, and governors of several prefectures including Huai An and Peng Cheng.

# 二十.
# 唐钱徽宅

## Site of Residence
## of Qian Hui of the Tang Dynasty

封疆万里等闲开，一日千金不吝财。持语无名求贡者，何须更入右银台。

唐代李复言《续玄怪录》载："殿中侍御史钱方义，故华州刺史吏部尚书徽之子。宝历初，独居常乐第。"可见，常乐坊中曾有钱徽及其子钱方义的宅第。

钱徽（755—829 年），字蔚章，吴郡（今浙江湖州）人。钱徽的父亲钱起是唐代著名诗人、"大历十才子"之一，官至尚书郎。钱徽于唐德宗贞元初年（785 年）进士及第，历经德宗、宪宗、穆宗、敬宗、文宗等朝，官拜中书舍人、华州刺史、翰林学士、礼部侍郎等。唐文宗太和二年（828 年）秋，钱徽以病辞官，授吏部尚书，次年三月卒，享年七十五岁。钱徽有两个儿子，一个是钱可复，另外一个就是《续玄怪录》中提到的钱方义。

钱徽是一个很正直的人。他曾任作礼部侍郎，知贡举。刑部侍郎杨凭用书画贿赂宰相段文昌，希望让他的儿子杨浑之及第。段文昌于是将此事托付给钱徽。同时，翰林学士李绅也向钱徽请托举子周汉滨。结果，钱徽并没有如二人所愿，秉持自己的原则，录取了杨殷士、苏巢等人。段文昌很生气，就向皇帝告状，说钱徽以权谋私。结果，皇帝让中书舍人王起、主客郎中知制诰白居易重新举行考试。钱徽因此事，被贬为江州刺史。这时候有人建议钱徽把段文昌、李绅向他请托的书信拿出来给皇帝看，一定可以洗清冤屈。不料钱徽却拒绝了这一建议，他说："如果我没有什么可以惭愧的地方，那么对于我来说，就不必计较得失。怎么可以用私人书信来作证明呢？"竟让家人把段、李二人的书信烧掉。大家十分佩服，都觉得钱徽是一个正直忠厚之人。

Qian Hui (755 - 829) was son of Qian Qi, a famous poet in the Tang dynasty, who recognized as one of the "Ten Talents" during the reign of Emperor Daizong of Tang. Qian Hui passed the imperial examination and attained the degree title of Jinshi when he was 30 years old. And then he served in the imperial court drafting and announcing an edict, and also as Governor of Huazhou Prefecture, Scholar in Hanlin Academy, and Deputy Minister of Rites through the reigns of Emperor Dezong, Xianzong, Muzong, Jingzong, and Wenzong.

Qian Hui was a person of integrity. When he once worked as Deputy Minister of Rites, he was responsible for enrolling candidates of Jinshi. Qian enrolled the candidates strictly based on standards and was framed for refusing to accept bribes. Qian was a clean and honest official. He repeatedly wrote memorials to the throne to denounce the trend of local officials presenting properties to the court and demanded that the court stop accepting tribute. People all admired Qian Hui and thought he was honest and upright.

# 二十一.
# 唐姚合宅

## Site of Residence
## of Yao He of the Tang Dynasty

旧客常乐坊，井水浊而咸。新屋新昌里，井泉清而甘。

姚合（约 779—855 年），唐代著名诗人，陕州（今河南陕县）人。姚合是宰相姚崇的曾侄孙，元和十一年（816 年）登进士第，授武功主簿，历任监察御史、杭州刺史、刑部郎中、给事中等职，终秘书少监。姚合世称"姚武功"，与当时的著名诗人刘禹锡、李绅、张籍、王建等人均有往来唱酬。其诗自成一体，善于书写萧条景物和"吏隐"生活，世称"武功体"。又因其诗风格与贾岛相近，被合称"姚贾"。

姚合有《新昌里》一诗云："旧客常乐坊，井水浊而咸。新屋新昌里，井泉清而甘。"可见，姚合先在常乐坊居住，后来搬到了新昌坊。

Yao He (779 - 855) was a famous poet of the middle Tang dynasty. In 816, he attained the degree title of Jinshi and was appointed as an official in charge of local books and classics in Wu Gong County, so he was often called "Yao Wugong". He served at several important positions such as an official in the Supervisory Department, the Governor of Hang Zhou, an official in Ministry of Penalty, and director of a department managing state books and classics, etc. Being a poet, Yao had a close contact with other famous poets of his time including Liu Yuxi, Li Shen, Zhang Ji, and Wang Jian. His poems featured describing desolate scenery and official seclusion, known as "Wugong Style".

# 二十二.
## 唐尉迟将军宅
## 王麻奴寓所

Site of Residence of General Yuchi Qing
of the Tang Dynasty
Site of the Flat of Wang Manu
of the Tang Dynasty

南山截竹为觱篥，此乐本自龟兹出。流传汉地曲转奇，凉州胡人为我吹。

唐德宗（742—805 年）时，常乐坊里住着一位吹觱篥的高手，叫尉迟青。尉迟青是西域于阗国人，官至将军。觱篥是传自西域龟兹国的乐器，有点像胡笳。

大历年间（766—779 年），幽州还有一位吹觱篥的高手，叫王麻奴。王麻奴架子很大，一般人想请他吹觱篥是请不动的。有一次，一个姓卢的官员要奉调入京，在送别酒宴上想请王麻奴吹奏一曲。王麻奴不肯。这位姓卢的官员很生气，就说：“你的才艺没有什么值得称道的，京城有一位尉迟将军，他的觱篥功夫才是冠绝古今呢！”王麻奴很不服气，当下决定进京与尉迟将军比试吹觱篥的功夫。王麻奴来到京城后，打听尉迟将军的住处，得知他住在常乐坊。于是王麻奴也在常乐坊尉迟将军府第的旁边租了一间房子，从早到晚吹奏觱篥，想引起尉迟青的注意。尉迟青经常路过王麻奴的门前，却假装没听到觱篥声。王麻奴很无奈，只好买通了尉迟青家的守门人，亲自登门求见。两人叙礼完毕，尉迟青请王麻奴吹奏一曲。王麻奴就吹奏了一曲西域乐曲，吹完后累得汗流浃背。这时，尉迟青拿起觱篥，吹奏了同一曲调，却吹得十分轻松，而且更加好听。王麻奴这才心悦诚服，拜而求教。

During the reign of Emperor Dezong of Tang (742 – 805), there lived a man named Yuchi Qing at Changle Square, who was a master of playing the Bili (an ancient bamboo pipe with a reed mouth piece), a musical instrument from Qiuci of the western regions. A man named Wang Manu, also an expert of playing the Bili, was quite arrogant and decided to compete with Yuchi when he heard of the general. Then Wang rented a flat near the General's residence at Changle Square and played the Bili day and night, trying to arouse the General's attention. With the General taking no notice of him, Wang had no choice but to bribe General Yuchi's porters for a visit. After greetings, the General invited Wang to play the Bili first. Wang Manu played a piece of music in western regions, and was sweating like a pig. But General Yuzhi, on the contrary, quite easily finished the same piece of music, more melodious. Totally convinced, Wang Manu hurried to the master for advice sincerely.

# 其他历史人物

　　在道政、常乐二坊中，除了上述人物和故事，很多其他的历史人物也曾生活在这里。根据清代徐松《唐两京城坊考》，当代学者杨鸿年《隋唐两京坊里谱》、李健超《增订唐两京城坊考》等古今学者的考证，还有以下历史人物曾在这片土地上生活过。

In addition to those mentioned above, there were dozens of other historical figures living in Daozheng Square and Changle Square, whose stories are to be explored. The list of the names were based on the studies by Xu Song in the Qing dynasty, and two contemporary scholars Yang Hongnian and Li Jianchao.

唐右神策军押衙朝散大夫襄王府咨议参军上柱国何少直

唐晋绛慈隰等州观察使试秘书省校书郎崔隋

唐右神策军同正将仕武将军守左金吾卫大将军员外置同正员上柱国赐紫金鱼袋李文政

唐朝散郎前行太史监灵台郎郭元诚

唐通议大夫行仪王府司马上柱国武阳郡开国公韦英

唐高祖李渊之女长沙公主

唐辅国大将军兼左骁卫将军御史中丞马实

唐太子太师浑释之

唐兖州都督于知微

唐光州长史检校国子祭酒上柱国边诫

唐左卫翊府左郎将兼监察御史包筠

唐幽州节度使衙前兵马使王承宗妻李元素

唐宣武军节度使押衙兼右器仗都头兵马使银青光禄大夫检校太子宾客上柱国李太均

**134**

唐镇国大将军王荣

唐国子监太学生武骑尉崔韶

唐上柱国宋感

唐文林郎杜俨

唐博士太子中舍人上柱国刘浚

唐左金吾卫将军襄武县开国男赵文皎

唐朝请大夫行尚书考功员外郎邵昃

唐京兆府长安县尉韦最

唐宁远将军守右司御率上柱国张令晖

唐左龙武将军梁约

唐太中大夫行殿中省尚药奉御孙嘉宾

唐朝散大夫太子左赞善大夫李咄

唐礼部尚书左龙武将军统军赠尚书左仆射史旻

# 后记

## Final Words
## to the Campus Tours

送你一个仙交大，一个带得走、有韵味、有想象力的交大。

如果，你走过这里的一天，触动你的，也许是她的颜值和声名；如果，你陪伴这里的四年，触动你的，也许是她的温暖和青春；如果，你了解这里的一甲子，触动你的，也许是她的情怀和坚毅；如果，你感受这里的两千年，触动你的，也许是她的博大和包容。

本书所展现的，不过冰山一角尔尔。

喜欢仙交大吗？想接着欣赏雁塔校区、曲江校区和中国西部科技创新港的风姿吗？欢迎给我们留言。你愿意分享自己在这片土地上的记忆和藏宝吗？请与我们联系，我们的邮箱：chl@xjtu.edu.cn。我们会以展览、出版物等形式陆续推出"穿越交大"系列。

我们，在这里等你。

一群热爱交大的交大人
2021 年 5 月 20 日
于兴庆校区

This is Xi'an Jiaotong University, or *Xian Jiaoda*, the Heavenly Palace — a university with a profound history and cultural heritage.

If you stay here for one day, you might be impressed by its beauty and reputation; If you stay here for four years, you might be impressed by its youth and vitality; If you learn about its history of the Western Relocation, you will be touched by its patriotism, dedication, responsibility and perseverance; If you learn about what happened on this land over two thousand years ago, you will be inspired by its broadness and inclusiveness.

The stories of XJTU introduced in this book are just the tip of the iceberg.

Are you fond of XJTU? If you want to learn more about each particular campus — Yanta Campus, Xingqing Campus, Qujiang Campus and iHarbour Campus, welcome to contact us.

If you'd like to share with others your stories with XJTU, please contact us. Your stories and ideas will be collected in the series of "The Tour Guide to Xi'an Jiaotong University" in the forms of exhibitions and publications.

E-mail: chl@xjtu.edu.cn

We are waiting for you at XJTU.

XJTUers
May 20,2021
On Xingqing Campus